GOLD MEDAL PHYSICS

GOLD MEDAL PHYSICS

The Science of Sports

JOHN ERIC GOFF

The Johns Hopkins University Press
Baltimore

© 2010 The Johns Hopkins University Press
All rights reserved. Published 2010
Printed in the United States of America on acid-free paper
9 8 7 6 5 4 3

The Johns Hopkins University Press
2715 North Charles Street
Baltimore, Maryland 21218-4363
www.press.jhu.edu

Library of Congress Control Number: 2009922669

ISBN-13: 978-0-8018-9321-6 (hc)
ISBN-10: 0-8018-9321-6 (hc)
ISBN-13: 978-0-8018-9322-3 (pbk)
ISBN-10: 0-8018-9322-4 (pbk)

A catalog record for this book is available from the British Library.

Special discounts are available for bulk purchases of this book. For more information, please contact Special Sales at 410-516-6936 or specialsales@press.jhu.edu.

The Johns Hopkins University Press uses environmentally friendly book materials, including recycled text paper that is composed of at least 30 percent post-consumer waste, whenever possible.

To Susan,
Emily, and Abby

CONTENTS

PREFACE

You got to be a man to play baseball for a living,
but you got to have a lot of little boy in you, too.
—ROY CAMPANELLA

I love that sentiment from Mr. Campanella, one of the finest catchers to have played the game.* Baseball, like all sports, is a game. Few things are as pure as the revelry of children playing a game, whether it be a sport or something else. Part of that joy stems, I believe, from the sheer wonderment of the unknown. When I played baseball as a child, I loved the "What will happen next?" feeling I held inside me.

That feeling applies equally to science. Who among us epitomize the essence of a scientist? Children. Unfettered by a self-awareness of their naïveté, children ask "Why?" They want to know why the sky is blue, why cars have wheels, why the hot dogs they get at the ballpark are called "hot dogs." As a parent, I know of the seemingly insatiable mental appetite of children. What's certain is that we've all been there. No matter your age now, there was a time when you clamored to *know*, with unquenchable curiosity. That's how scientists behave. We simply want to *know* how the world works. When I learn something new while pursuing my work as a physicist, I feel just as giddy as I did when I got a base hit playing Little League.

This book combines my love of sports with my passion for science. The wide world of sports is wide indeed. But even though I discuss sports

---

* Quote taken from *Baseball Legends of All Time*, Publications International, Ltd. (1994), p. 42.

played all over the world, I certainly cannot invoke them all in this small ballpark. What I hope to do is to show you how one may view the world of sports through the lens of physics. The sports events I have picked are either ones I watched with great enthusiasm or ones that fascinated me when I read about them.

I've written this book for someone with a liking for sports and at least a passing interest in science. Nowhere do I push the physics envelope too far beyond what a first-year university physics course covers, and there are plenty of places in the book where you may want to venture beyond what I explore. The book is thus not meant to be a textbook for a "Physics of Sports" course, though I believe it could be used in such a course, at least for supplemental reading and for project ideas. If you are a sports fan who can tap that little kid within you, the one who was so good at asking how the world works, this book is for you.

Many people helped bring this book into existence. I wish first to thank Trevor Lipscombe, my editor at the Johns Hopkins University Press. I never dreamed of writing such a book until Trevor approached me with the idea. He taught me much about writing, and I hope I've eliminated as much "purple prose" as possible. Also at the Johns Hopkins University Press, Julie McCarthy, Kim Johnson, Robin Rennison, Claire McCabe Tamberino, Kathy Alexander, Brendan Coyne, Greg Nicholl, Thomas Maszczenski, and Erin Cosyn helped me at various points in the publishing process. Bill Carver did a phenomenal job copy editing the book. I probably learned as much about the English language from Bill as I learned about sports while writing this book.

I owe thanks to many Lynchburg College faculty and students. My physics colleague, Julius Sigler, read many chapters and offered valuable feedback. I benefited from insightful conversations with Allison Jablonski, John Styrsky, George Schuppin, Rich Burke, and Loretta Dorn. I thank Robert White for doing some English/French translations for me. Chapters 4 and 7 would not have been possible without the great ideas offered by two of my former research students, Ben Hannas and Brandon Cook. I thank Jack Toms for the track and field conversations we had, and I am grateful to Chris Yeager and the Lynchburg College men's soccer team for allowing me to disrupt one of their practices. I thank Rory Lee-Washington for an interesting chat about discus throwing. Two graphic artists did fine work for me. Rebekah Hoskins created Figures 2.1 and 2.5. Kenny Wright created Figures 2.3, 7.4, 7.6, and 7.8. Thanks to Bryan Atchison for some useful suggestions for Chapter 1. Robin Berrington provided me with a wonderful start to my sumo research. Jay and Mandy Gerlach and Jarrett Gerlach have made the material I cover in Chapter 10 great fun.

Finally, nothing I do is possible without my family. I thank my beautiful daughters, Emily and Abby, for their love and for reminding me that the best scientists are children. My wife, Susan, helped me in every way imaginable. She read many chapter drafts, offered unparalleled advice, and was amazingly patient when I needed time at the keyboard. But despite all the people who have read and edited the book, it is quite possible that something has fallen through the cracks. Any remaining errors are solely my responsibility.

GOLD MEDAL PHYSICS

1

The Pre-Game Show

Crack! For some people, that sound of a wooden bat just crushing a baseball is one of the sweetest sounds they may hear. It evokes memories of childhood, when they honestly thought they could one day make the big leagues. If you have never played baseball, that sound may not tickle your mind any more than, say, a car's backfire might. We all experience the world in different ways. The point is, the more we learn about our world, the richer our experiences can be.

For example, my wife and I attended a Cleveland Indians game in the summer of 2001. We sat in the right-field section of Cleveland's wonderful Jacobs Field. It was a perfect evening for baseball. I had played baseball as a kid and have seen way too many ballgames to count. The sights and sounds of the game are as familiar to me as just about anything. In contrast, my wife could count on one hand the number of baseball games she had seen.

Early in the game, a batter hit a lazy fly ball to right field, what a baseball aficionado would call a "can of corn." As the fielder lobbed the ball back to the infield, my wife asked, "Isn't it weird how the ball came off the bat before I heard the crack?" It dawned on me at that moment just how differently people experience the world. For I am a physicist and my wife, Susan, is a Japanese translator.

A couple of months before the Indians game, she and I had spent two weeks in Japan. I had never been out of the United States, except to see the Canadian side of Niagara Falls. Susan, for her part, had been to various parts of Europe and Asia and had lived in Japan for about five years. When I got off the thirteen-hour plane ride, I was no better prepared for what I would see than the deer caught in the headlights. Susan was right at home. In the United States, I hardly notice gas stations, restaurants, malls, hotels, houses, doctors' offices, etc. They are all just familiar background. But when I was in Japan, I had little idea what I was seeing when I walked down a street. Fortunately for me, Susan moved effortlessly through the streets and the subways. Words there, written in characters strange to me, were utterly obvious to my wife. Suffice it to say, she saw the world around her differently from the way I saw it. Just as I asked her what was listed on a restaurant menu, she asked me about the sound of a hit baseball.

The one place in Japan where I felt right at home was at a baseball game. While in Yokohama, we saw the hometown Bay Stars take on the Yakult Swallows. Japanese baseball is a bit different from what I was accustomed to in the States. The players and stadia are smaller than their American counterparts, and there is a greater emphasis on "small ball" (bunting, stealing, and sacrificing). Similarly, pitchers rely more on "junk" (curveballs, sliders, screwballs, gyroballs, and the like) than on power. Each team had a band stationed in the outfield seats. Every half inning, they took turns playing their respective team songs, and the fans sang right along. Aside from these differences, and the fact that fans drank sake and ate fish, everything else was essentially the same. Baseball is baseball.

We can all learn things that make the way we experience the world much richer. I would have been a more intrepid traveler had I known some Japanese, and Susan would have thought more about the difference in speed between sound and light had she known a little physics. I hope that as you read the pages in this book, you will gain richer experiences from the world of sports. Perhaps you will be watching a sporting event in the future, and some aspect of it will be clearer to you than it had been in years past. That would be gratifying.

So, why *do* we hear the crack of a baseball from a bat a split second after we see the ball leave the bat? Because the speed of sound and the speed of light are different. Most objects we see are visible only because light from some other source bounces off of the objects and into our eyes.[1] The sun is typically the most dominant source of light during daytime. Halogen lights or incandescent light bulbs provide light when we are inside, or when we need help seeing things at night. There are certainly other light

---

[1] I am greatly oversimplifying how light interacts with matter when I use the verb "bounces."

sources; the point, however, is that when we see a ball make contact with a bat, we are seeing the assemblage of light patterns that bounced off of them.

Sound must also travel to us if we are to hear it. Clap your hands together and you hear a sound. Some of the energy in the motion of your hands is transferred to an energetic motion of the air molecules around your hands. The sound wave produced eventually reaches us and transfers some of its energy to our ears. Unlike light, which travels quite well in an empty vacuum, sound needs a substance through which to travel. That substance, for most of the sounds we hear, is air. (If you have ever had your head underwater in a swimming pool, you have probably heard sounds there, too.) Light in fact achieves its greatest speed in a vacuum, whereas sound cannot travel at all in a vacuum, because there is no supporting substance. And whereas light speed drops by about 25% when passing from air to water, sound speed actually *increases*, by more than a factor of *four*, when it moves from air to water.

At the instant the bat hits the baseball, sound waves and reflected light waves leave the scene of the collision. In air, light travels roughly 300,000,000 meters per second[2] or 186,000 miles per second. At such a great speed, light could circle the Earth's equator about seven-and-a-half times in a single second. *Nothing* travels faster than light; light speed is the universe's speed limit. Sound, on the other hand, travels far slower. In warm evening air,[3] at sea level, sound travels about 345 m/s, or roughly 772 mph (miles per hour), or 1242 kph (kilometers per hour). For comparison, in the time it takes light to circle the globe[4] once, sound travels about half the length of a 100-yard football field.

Susan and I were probably sitting about 450 feet (137 m) from home plate. That meant that it took light about half a *microsecond* to reach us. A microsecond is 10^{-6} s, one millionth of a second, a completely

---

[2] I'll write such numbers using *scientific notation* and abbreviations for the units. Thus, light speed is roughly 3.0×10^8 m/s (meters per second) in air.

[3] At temperatures familiar to us, sound speed in air increases as the temperature increases, by about 0.6 m/s (1.3 mph) for each °C.

[4] Light would not naturally do such a thing. Even if the light did not bump into a bunch of stuff at the ballpark, it would travel in a straight line. Light *could* circle the globe using something like a fiber-optic cable. In that case, one would have to worry about extremely low light *intensity* after a few trips around the world.

imperceptible passage of time for us.[5] The sound wave, for its part, needed about 0.4 second to reach us. That is roughly the same time a good major-league fastball takes to go from the pitcher's hand to the catcher's glove. Thus, because light travels almost a million times faster than sound, Susan and I saw the ball-bat collision before we heard it. This effect is of course not noticeable to the batter. Little more than a meter away, the sound from a ball-bat collision reaches the batter's ear in roughly five milliseconds, or 5×10^{-3} s. That time is about twenty times shorter than the time needed to blink. Even if light speed were infinite, the travel time for sound would still be too small for the batter to notice any difference between when the collision was seen and when it was heard. To convince you of this, the shutter time for a camera set to *250* is 4×10^{-3} s. Honeybees flap their wings every 5×10^{-3} s or so. You could no more perceive much during the time when a camera clicks than you could notice each flap of a honeybee's wing. When the light from the ball-bat collision reached our eyes, out past right field, the sound wave had not yet traveled even two-tenths of a *millimeter*. Put another way, when the visual information reached our eyes that the collision had taken place, the information carrying the sound of the collision had traveled only about the width of two human hairs. The light could have traveled one-third of the way to the Moon in the time needed for the sound wave to reach our ears. My wife and I, and everyone beyond, in the nosebleed seats, perceived the time delay, whereas those in the expensive seats did not.

Although we humans might have our curiosity piqued by the idea of a perceptible travel time for sound at a baseball game, the survival of some animals relies upon the time delay of sound travel. Many types of bats use *echolocation*[6] to find their way in the total blackness of night. Aside from using sound waves to keep from flying into trees, bats use echolocation to find food such as insects fluttering about in the night.

Some whales and dolphins are also known to use echolocation. The internal sonar of these animals is incredibly sophisticated. A killer whale (orca) can send out a wide beam of sound. If some of that sound bounces

---

[5] AM medium-wave radio broadcast wavelengths of 300 m have frequencies of 10^6 cycles per second (one megahertz, or 1 MHz). The time for each oscillation is thus 10^{-6} s (1 MHz is therefore the same as 1 cycle per microsecond), a time far too short for us to notice.

[6] Another term for echolocation is *biosonar*.

off prey to the whale's left, the whale detects the echo of that sound in its left ear before the echo reaches the right ear. Even small differences between the intensity of sound reaching one ear and the intensity of the same sound when it reaches the other ear are noticed. Echolocation is extremely important for mammals living and hunting where there is not much light, whether at night or in the ocean.

For us humans, who rely so much on vision, echolocation is not vital to our survival. We can, however, have fun with echolocation. Instead of using a known value of sound speed to determine a distance, try using a known distance to determine the speed of sound. The next time you find yourself at a known distance from a barn, a mountain cliff, or even an isolated building, scream a great word like "Baseball!" Use your wristwatch to determine how long it takes the echo to reach you.[7] Suppose you are two-tenths of a mile from a barn and you scream. You then time the echo and find that your scream took about 2 seconds to get back to you. Given that the sound traveled a total round-trip distance of four-tenths of a mile, you would divide distance traveled by time[8] and get about 720 mph. That's not too bad, considering that you are standing in a cow pasture screaming at a barn and staring at an ordinary wristwatch.

Pumped up by the excitement of obtaining an estimate of sound speed, you might be tempted to try a similar experiment to determine light speed. Suppose you put a big mirror on the barn at which you just screamed, a mirror aimed at the position where you screamed. Then return to your screaming position with a strong flashlight or, even better, a pen laser. Look at the mirror with a pair of binoculars and then flip on the pen laser. You will find that any attempt to measure the flight time of light with your wristwatch will fail. Galileo tried a similar experiment using a couple of lamps and the known distance between the tops of adjacent hills.[9]

---

[7] As with the ball-bat collision, this echo experiment is more fruitful if you are at least a couple of hundred meters (one or two tenths of a mile) away from the object at which you are screaming.

[8] Remember to convert the time units from seconds to hours. There are 3600 seconds in one hour. Thus, $2\,\text{s} = 5.\bar{5} \times 10^{-4}$ hours, where the bar over the five means that 5 repeats forever after the decimal place.

[9] Galileo Galilei (1564–1642) performed experiments that laid the groundwork for modern science. An account of his light speed experiment can be found in his *Dialogues Concerning Two New Sciences* (1638). In 1954, Dover reissued a 1914 translation of this work by Henry Crew and Alfonso de Salvio.

One person on one hilltop flipped on his lamp and the other person was supposed to flip on his lamp when the first lamp's light was seen. Galileo thought that light's travel time could be timed; all that he could determine, however, was that light speed must be some incredibly large number, because he found that all measured reaction times could be explained by human reaction time.

Modern experiments have determined the speed of light. We now use an exact value of the speed of light in vacuum, namely 299,792,458 m/s, to define the meter, which is the distance light travels in one 299,792,458th of a second. One artifact of Galileo's experiment is still true today—the speed of light can be found *only* in a round-trip experiment.

How to Do Physics

To answer my wife's bat-and-ball question and estimate the sound speed on a farm, I made several assumptions about how the world works. Some of those assumptions are so ingrained that they hardly seem like assumptions at all. For example, I calculated the time it took for either light or sound to go from the ball-bat collision point to Susan's ear using the following equation:

$$\text{time} = \frac{\text{distance}}{\text{speed}}. \tag{1.1}$$

This is a simple model, one that is used often. Suppose you wanted to know how long it would take you to drive from Yankee Stadium to Fenway Park, for example. New York and Boston are roughly 200 miles apart. Now suppose that with all the traffic, construction, tolls, and other such things that slow us down, the best you could do on your trip is to average about 50 mph. Equation (1.1) estimates your time of travel as

$$\text{time} = \frac{200 \text{ miles}}{50 \text{ miles/hour}} = 4 \text{ hours}. \tag{1.2}$$

This model is simplistic, to be sure; this is, however, what physicists do when we try to describe how the world works.

There are a few different ways that physicists study the world. We go out into the world and make observations. We do experiments and make measurements; we make many measurements and try to reproduce (or "replicate") experiments that we (or others) have made in the past. To measure the time it takes sound from a ball-bat collision to reach the right-field seats, for example, I could set up a couple of sound sensors, one near the batter and one near where I would be sitting. Then I would measure the difference in times recorded by the two sensors. But because this little experiment would be difficult to do during a game, I would have to set something up in a more controlled environment.

Suppose you do the experiment. You now have a number for the time it takes sound to go from a location close to home plate to a specific seat in the right-field section of a ballpark. So far, so good. But what if you go to another game and you sit right behind home plate? You won't notice the time delay between seeing the bat hit the ball and hearing the "crack." To determine how long it will take the collision sound to reach your ears, you don't need to set up another experiment. You use a model that allows you to predict the sound's travel time. If your seat behind home plate was ten times closer to the plate than your seat way out in right field, you could make use of equation (1.1) and determine that the sound would take only one-tenth the travel time to reach the seat behind home plate as it would to reach the seat in right field. The shorter sound travel time would in fact be only 0.04 second. A human blink takes about two-and-a-half times longer than that, so you are probably unable to notice much delay in that 0.04-second time frame. The real power in a model is thus its ability to make predictions.

Another way physicists study the world is by creating an intellectual model and seeing if it predicts what will happen in some experiment conducted in the real world. The idea is to use known laws of physics (or perhaps create new ones) and determine the consequences of those laws. The models we create are thus useful only if they match, to a level of accuracy we can detect, what we see in the real world. We can also use computers to perform simulations of the world, using models we program. The results of those simulations can then be compared with experimental data.

What we hope to do with all the models we make is to learn how the world works. Sound is a longitudinal wave. I neither proved that, nor used it, to determine a time-of-flight estimate. Experiments have been performed that convince us that sound is a wave, and we feel confident that one of the fundamental properties of sound is that it *is* a wave. We could go much deeper and try to understand whether our model actually helps us uncover what sound actually *is*. But that's for a discussion with a philosopher.

One important phrase in doing physics is "level of accuracy." When we make models, we concern ourselves with how accurately we wish to know something. In our time-of-flight example of sound versus light, I stated only one-digit accuracy in these times.[10] The distance from home plate to the right-field seats was merely an estimate. The speed of sound I used was reasonably good. The speed of sound, though, depends on the air's density and temperature. To know the sound's flight time to, say, four-digit accuracy, I would need a far more detailed model than the one I employed. I would need to know the air density and the air temperature at the moment of the test that evening to get a better estimate of the sound speed. I might even need to know something of the relative humidity. The time delay at a Florida Marlins game in August would be different from the delay at a Chicago White Sox game in the post season. I could easily have addressed the *qualitative* aspect of my wife's question by saying, "Light speed is far greater than sound speed, and what you noticed is the time delay between seeing the ball get hit and hearing the crack of the bat." That is probably what I told Susan when she first posed her question. Had she wanted more details, I might have estimated the numbers you have read here. If she wanted even more detail, I would have had to remind myself of the empirical[11] equation that determines the speed of sound for a given temperature and altitude.

Sports and Beyond

Physicists love to learn how things work. We study the goings on inside the protons and neutrons that make up atomic nuclei; we study how the universe was formed. We also study a great deal of what lies between those

---

[10] Some things need not be known accurately. Do you care if your indicated oven temperatures are accurate only to three (or even *two*) digits? Other things need more accuracy. In 2000, Texas Governor George W. Bush beat Vice President Al Gore in Florida by a count of 2,912,790 to 2,912,253, a difference of only 537 votes. In that case, a minimum of *five*-digit accuracy was needed.

[11] By *empirical*, I mean using actual experimental observations instead of theoretical derivations.

two realms. The protons and neutrons that make up atoms have a size of roughly 10^{-15} m, whereas the universe is thought to be at least 10^{26} m wide. For those of you scoring at home, physicists study entities varying more than *forty* orders of magnitude (i.e. powers of ten) in size. What allows us to be so flexible in what we study is the inflexibility in the laws of physics. In other words, we apply the same laws of physics to whatever we study. Those laws, though, sometimes need modification. To study atoms, for example, we need to use something better than the laws of motion that Newton gave us.[12] Newton's laws work really, really well for the macroscopic world, and are used all the time in studying sports.

We cannot, however, make use of those laws in the atomic world; there, we need *quantum mechanics*. There are other modifications to Newton's laws. For example, if an object moves at some appreciable fraction of the speed of light, we need *special relativity*, which reduces to Newtonian dynamics only for speeds much less than light speed. If we study objects with large mass, such as galaxies, we may even need *general relativity*. Although the sumo combatants we will be examining later in the book are pretty massive, we will be safe neglecting effects from general relativity. All of the aforementioned modifications are well beyond the scope of this book, for they are not directly observable to a sports fan's naked eye. Even though the sports world we see does not require such modifications for its description, there is an entire industry that strives to make better materials for golf clubs, golf shirts and shorts, biking shirts and shorts, etc. The technologies needed in making new materials rely heavily on molecular physics and chemistry, in which quantum mechanics plays a crucial role.

There are a couple of ways to tackle the world of sports. You could learn a bunch of physics and then examine the sports world with a reasonably full box of the scientific tools needed for an understanding. Or, you could jump right into the sports arena and try to understand what is going on, with a little help from a physicist. The latter is the route we shall take. To do so, we need some equations. After all, *cause* and *effect* are neatly stated in an equation. For example, in equation (1.1), above, if the speed

[12] Sir Isaac Newton (1642–1727) published his *Principia Mathematica* in 1687. The laws he set forth in that book form the basis of what we now call *classical mechanics*. It was not until the end of the nineteenth century that scientists discovered domains in which Newton's laws did not hold.

is held fixed, doubling the distance (a cause) will double the time needed to cover that distance (an effect). I would rather write a few equations down than try to say it all in words.

After learning some physics in the context of sports, you may understand things beyond sports. As a child, were you ever told how to determine how far away lightning struck? After seeing a lightning flash, immediately begin to count. Really accurate counting that goes like, "One-thousand-one, one-thousand-two," and so forth! Every five of my counted "seconds" was said to represent 1 mile, and when I heard the thunder, I simply divided the number of counts by five, and that represented how many miles away the lightning struck. That childhood rule works for the same reason that there's a time delay in hearing the "crack" of a baseball bat in the nosebleed seats. It also works because light travels nearly a million times faster in air than sound does. So, let us forget about the light travel time, since it will always be about a million times smaller than the sound's travel time. Thunder occurs because a lightning discharge causes a rapid expansion in the air. One of those longitudinal waves I mentioned earlier then propagates away from the path of the lightning discharge. If the speed of sound is 772 mph and we use equation (1.1), we can determine the time it takes sound to travel 1 mile, as follows:

$$t = \frac{1 \text{ mile}}{772 \text{ miles/hour}} \simeq 0.0013 \text{ hours} \simeq 4.7 \text{ seconds.} \qquad (1.3)$$

The symbol "$\simeq$" means "approximately equal to." I use that symbol because I have rounded times to just two-digit accuracy. Sound travels 1 mile approximately every 5 seconds. So, a reasonable rule of thumb for determining how far away a lightning storm is is to count the time between seeing the lightning and hearing the thunder and then dividing the number of seconds by five. It's crude, but it's not too bad. If you wanted greater accuracy, like those who work in meteorological and atmospheric science, you would need to do a little better than simple voice counting.

Toy Models

"Toy" models are used by physicists to get a feel for what is happening in the system they're examining. Trying to model, say, a human being running is exceedingly hard. We can qualitatively predict when a sprinter will cross the finish line. But if we wish to investigate the biomechanics of running, we need something more. So-called "ball-and-stick" models are useful, where joints are assumed to be points (or balls) and bones are assumed to be rigid rods (or sticks).

You have most likely played with a spring at some point in your life. If you never had a Slinky, you have surely noticed springs in objects like flashlights that require batteries. A *spring* is one of the toy models we use, in this case to make connections with a whole host of phenomena, such as light and sound. You might wonder what a spring has to do with light and sound. It turns out that the way a spring behaves, as long as it is not stretched too far,[13] is similar to the way the entities that bounce back and forth in a light or sound wave behave. Springs, light, and sound all exhibit this oscillatory behavior.

Figure 1.1 shows an ordinary wooden ruler stuck in my kitchen table. It is in an *equilibrium* configuration, doing nothing and going nowhere. In Figure 1.2, my finger pushes the ruler, displacing it from its equilibrium configuration. If I do not push the ruler too far before I let go of it, it returns to its equilibrium position, perhaps pursuing a few oscillations on the way to resuming that position.[14] A great number of things behave like this. Push on your stomach and it bounces right back. Rock your computer monitor up a little and it will probably come back to where it originally sat. Engineers are well aware that high winds cause tall bridges and buildings to "give" a little, to "sway" a bit. Luckily for us, bridges and buildings usually return to their equilibrium positions after the winds settle down. If we understand how a simple foot ruler behaves when it feels a force (or stress), we are a long way to understanding how a diving board works.

We can describe all of the aforementioned examples in a qualitative way as springs, because if you stop pulling on a spring it will return to its original (or equilibrium) position. Because so much of nature behaves in ways similar to those of springs, the spring is one of the ubiquitous

---

[13] Stretching a spring too far will exceed what is called the *elastic limit* of the material.

[14] Those who have had some physics will be thinking about *damped oscillators* at this point.

Figure 1.1. Configuration of a ruler at equilibrium.

toy models we study in physics courses.[15] We will return to the spring in more detail when we look at diving.

Ready for Sports!

I love to play and watch sports. I also enjoy physics, and I am continually fascinated by the fact that the wide world of sports offers a seemingly endless supply of great examples of physics. Though we probably derive most of the pleasure we experience with sports by playing or watching

---

[15] One significant feature of a simple spring is that the spring's restoring force, F, is proportional to the distance, x, that the spring has been displaced from equilibrium (i.e. $F \propto x$). A spring that has $F \propto x$ obeys *Hooke's law*. The key point here is that a simple spring has a restoring force that is *linear* in its displacement. Thus, a great deal of what we call *linear physics* can be modeled with springs. *Nonlinear* models use force equations with higher-order terms in x, such as x^2, x^3, etc. One could also include terms with trigonometric functions, logarithms, etc.

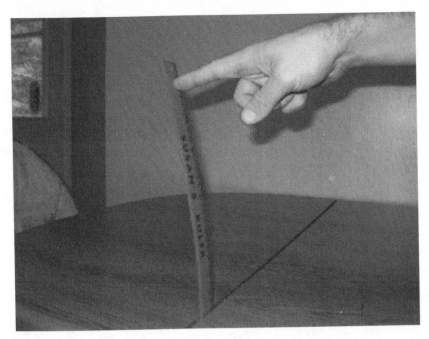

Figure 1.2. My finger putting stress on a ruler.

them, or simply by cheering for our favorite teams, I believe we can have fun studying them. Watching a sporting event and realizing that you have a good idea why (and how) something just happened is deeply satisfying.

There is perhaps also a downside to learning some physics in the context of sports. Hearing things like "He hung in the air on that dunk!" and "That pitch fell off the table just before hitting the catcher's mitt!" will provoke a whimsical smile instead of a rousing endorsement. We will not actually examine a basketball player's "hang time," nor will we examine the "wicked" curve balls thrown by a pitcher. We will, however, explore the physics that demonstrates the errors embodied in those two comments. Beyond sports, you may even find that watching movies will be different. One of my favorites is "Star Wars"; but my mind spends too much time noting what is physically impossible in the movie. There are explosions in outer space that make noise;[16] there are engines running while space

[16] Is there a *substance* out there through which sound can travel?

ships are in deep-space travel;[17] and there is the dreaded "warp" speed where ships go faster than the speed of light.[18]

I hope that as you read these pages and then play and watch sports, your mind will be a little more attuned to some of the neat stuff that goes on in the sporting world. With the conclusion of the pre-game show, we turn our attention now to the famous ending of the 1982 Stanford-Cal football game.

[17] Do the folks driving the ships know about Newton's first law, which states that an object in motion at a constant velocity will remain that way unless acted upon by a net external force?

[18] Dare you tell Chewbacca that he was acting like a *tachyon*? A tachyon, a particle conjectured by some theoretical physicists, has the property of traveling faster than the speed of light. If Chawbacca *were* acting like a tachyon, he could not slow down to light speed or any slower!

2

It's Not Over Until the Trombone Player Goes Down

Vectors and How to Think Like a Physicist

"Oh, the band is out on the field!"

"The Bears have won! The Bears have won! Oh my God, the most amazing, sensational, traumatic, heart-rending, exciting finish in the history of college football!" Joe Starkey uttered those words exuberantly more than a quarter century ago. I did not watch "The Play" (as it is now known) live, but I have surely seen it dozens of times in my life. If you love college football, or even if you are a casual fan, you have seen the crazy end to the 1982 Stanford-Cal game at least once, even if you were not yet alive when it happened. Some sports plays are destined to live forever, and the Cal Bears' improbable victory over the Cardinal in the eighty-fifth edition of the "Big Game" is surely one of those immortal plays.

So, what is so special about "The Play"? We need to go back to November 20, 1982, and visit Memorial Stadium in Berkeley, California. Even though the rivalry between Stanford University and the University of California, Berkeley, may not receive the spotlight as brightly as the yearly matchups between, say, Ohio State and Michigan, or Alabama and Auburn, it is still intense.[1] Some of the Stanford players who were part of the famous 1982 game will not discuss it. I have spoken to Stanford

---

[1] As someone who did his undergraduate work at Vanderbilt University, I know exactly where I was on the Saturday evening of November 19, 2005, when I heard on the radio that my beloved Commodores had beaten the University of Tennessee in Knoxville, 28–24. That was our first win over the Volunteers in 23 years.

alumni who refuse to acknowledge that Cal really did win the game. "The Play" was the last play of the game; it won the game for Cal; there were *five* laterals during the play; and the Stanford marching band (one trombone player in particular) became part of the chaos that erupted on the field.

John Elway, who went on to greatness in the National Football League, was the senior quarterback for Stanford. He was Pac-10 Player of the Year that year; he led the nation with 24 touchdown passes and would be the runner-up to Georgia's Herschel Walker for the 1982 Heisman Trophy. But for all of Elway's personal achievements, he had never led Stanford to a bowl game during his first three years in Palo Alto. The game against Cal in 1982 was thus huge for Elway and Stanford because the Cardinal was 5-5 heading into the "Big Game": a win over Cal on that Saturday before Thanksgiving would probably have meant an invitation to the Hall of Fame Bowl.

The Bears were up 19–17 with just 53 seconds left on the clock, and Elway faced a 4th and 17 on his own 13-yard line. As he would do time and time again for the Denver Broncos, Elway drove his team down the field, hoping for a come-from-behind win. He first completed a 29-yard pass on that 4th and long, which got Stanford to their own 42-yard line. A 19-yard pass then got the Cardinal to the Bears' 39-yard line. Tailback Mike Dotterer had an impressive run of 21 yards, taking the ball all the way to the Cal 18-yard line. Dotterer then ran to his right for no gain, setting up a 35-yard field-goal attempt near the right hash mark. As soon as Dotterer was down, Elway signaled time out; it was Stanford's final time out. Unfortunately, Elway had called it with 8 seconds left on the clock. Kicker Mark Harmon nailed the 35-yard field goal, leaving 4 seconds on the clock. Had Elway waited 5 or 6 seconds after Dotterer was tackled, the field goal would have occurred when the clock hit double zero, and the game would have ended with a 20–19 Stanford win. Stanford players would have stormed the field and celebrated a huge come-from-behind win, a winning season, and a likely bowl bid. Elway would have been heralded as the game's hero and might have fared better against Herschel Walker for the Heisman.[2]

[2] Elway was 695 points shy of Walker for the 1982 Heisman Trophy. In all six regions of the country, Walker beat out Elway, who finished second in every region.

After Harmon's field goal, Stanford was penalized 15 yards for excessive celebrating. Thus, Harmon had to kick off from the Stanford 25-yard line[3] with 4 seconds left on the clock. Instead of kicking the ball deep and allowing Cal the opportunity to set up a good return, Stanford elected to "squib" the ball, kicking it short and low to the ground. This should allow the kicking team to get to the ball carrier quickly and make the game-ending tackle. The risk was that Cal already had a short field with which to work. With only 4 seconds left in the game, Cal had just two choices: get the ball out of bounds quickly after fielding it or take a knee. Either would have stopped the clock. That would leave one final "Hail Mary" for the end zone. Another option was for the Cal players to set up a great return. The odds of the Hail Mary working were slim, to be sure; the odds of the great return were even slimmer. Harmon's squib kick was fielded by Kevin Moen on the Cal 45-yard line. What happened next will live forever in college football lore.

"The Play" is diagrammed in Figure 2.1. Moen made a rough semicircular scramble to the Cal 48-yard line and then lateraled the football to Richard Rogers, who was standing at around the Cal 47-yard line, about midway between the hash mark and the sideline. Rogers gained only about a yard before he lateraled the ball straight backwards to Dwight Garner, who caught it between the Cal 43- and 44-yard lines. Garner then ran straight ahead until he was just about at midfield, where he was gang tackled by several Stanford players.[4] As he was going down, Garner pitched the ball to Rogers, who was right behind Garner at the Cal 48-yard line. The 4 seconds on the clock at the start of the play were gone by now, meaning the game would end if Stanford could tackle the Cal ball carrier

---

Walker had such a large edge in first-place votes (525 to 139) that it is doubtful that Elway would have overtaken Walker even if Stanford had beaten Cal on that fateful day.

[3] Without the penalty, Harmon would have kicked off from the 40-yard line. College football kickoffs were moved to the 35-yard line in 1986. They were moved to the 30-yard line in 2007 to increase the number of returns.

[4] Whether or not Garner was successfully tackled before he lateraled the ball is still the subject of contentious debate. I have watched the frame-by-frame sequence of "The Play" several times, and it appears to me that Garner was indeed *down* before the lateral happened. Sorry, Cal fans! It must have been difficult for the referees to make that call.

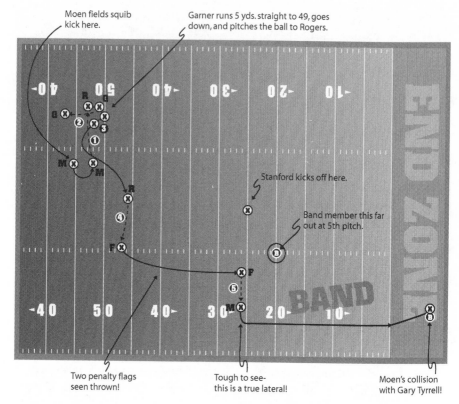

Moen fields squib kick here.

Garner runs 5 yds. straight to 49, goes down, and pitches the ball to Rogers.

Stanford kicks off here.

Band member this far out at 5th pitch.

Two penalty flags seen thrown!

Tough to see- this is a true lateral!

Moen's collision with Gary Tyrrell!

Figure 2.1. A schematic of "The Play." The laterals appear as dashed lines. The letters represent the last names of the relevant players (Harmon, Moen, Rogers, Garner, and Ford).

before that player could score. But they didn't. Rogers circled back into Cal territory and then burst forward to the center of the field at the Stanford 45-yard line. At that point, Rogers executed Cal's fourth lateral to Mariet Ford, who caught the ball on the run at the Stanford 47-yard line. Ford was just to the right of the right hash mark when he received the ball from Rogers. As Ford crossed the Stanford 40-yard line, a yellow flag was thrown, while another already rested on the turf. The refs apparently noticed that the Stanford marching band, its players assuming that the game was over, was on the playing field and the refs were thus penalizing Stanford. Ford, meanwhile, made it all the way to the Stanford 26-yard line before he pitched the ball to Moen, who had originally fielded the kickoff, and who was now between the Stanford 26- and 27-yard lines and about halfway between the right hash mark and the sideline.[5] As Moen caught the ball, a few members of the Stanford marching band could be seen near the right hash mark of the Stanford 20-yard line. Moen, with no more tacklers in sight, dashed into the end zone for the winning score and then collided with Stanford trombone player Gary Tyrrell, who was about 6 yards deep in the end zone at the time.

The crazy touchdown gave Cal the victory over Stanford that day, 25-20.[6] Stanford players, band members, and fans were stunned. Conversely, the Cal faithful were jubilant. Elway and the Cardinal had been denied a bowl bid,[7] and controversy over "The Play" would rage for years. What is undebatable, however, is that the end of the 1982 Stanford-Cal game had given college football history one of its most indelible moments.

Physics Time

Now that we have discussed "The Play," we can look at it again through the eyes of a physicist. The idea here is not to take anything away from the sheer enjoyment of witnessing a great sporting moment. Rather, we can use that moment to elucidate some of the basic ideas in physics. The wide world of sports offers many exciting and practical examples of physics in action. Armed with a little physics, we can then go back and understand more about the sporting moments we see.

To begin this back-and-forth between sports and physics, let's try to understand how Mark Harmon managed to get to the football, in the few steps he took before kicking it. Have you ever thought about how you

---

[5] This fifth and final lateral is also the subject of much debate. Many people believe that this last lateral was a forward pass and should have earned Cal a penalty, thus nullifying the final touchdown. Again, I watched this part of "The Play" frame by frame many times. Unlike the issue of whether or not Dwight Garner was down before he executed Cal's third lateral, I believe it is much more difficult to settle the issue of whether or not the final lateral was indeed a forward pass. It was probably too close to call.

[6] Cal did not need to kick the extra point.

[7] Though Elway's final collegiate game was a heart-rending loss, albeit a famous one, his final game as a professional was much better for him. He ended his career with the Denver Broncos by defeating the Atlanta Falcons in Super Bowl XXXIII on January 31, 1999. Elway was named Super Bowl MVP.

walk? Sure, you put one foot in front of the other. But, what is actually pushing you forward? Are *you* pushing you forward? If you believe that Harmon pushed himself forward in order to get to the football, ask yourself if anything would have been different if there had been snow and ice on the field.

Newton's third law[8] helps us understand walking. It says that if one object feels a force from a second object, called the "action," the second object feels a force from the first, called the "reaction," and that force is of the same magnitude as the first force, but it acts in the opposite direction. Note that although the terms "action" and "reaction" are often used with Newton's third law, the "action" force does not *precede* the "reaction" force. Two objects exerting forces on each other simultaneously feel a force. Also, I use the words "magnitude" and "direction" because physicists model forces using mathematical objects called *vectors*. The magnitude[9] of a *force vector* is the size of the force. In other words, if I push on your arm toward the east with 10 pounds[10] of force, the magnitude is "10 pounds" and the direction is "east." Note that your arm pushes on my hand with 10 pounds of force to the west. The implication of Newton's third law is that forces always come in pairs.

Figure 2.2 shows my foot pushing back on the ground. The ground pushes forward on my foot with the same magnitude of force, but in the opposite direction. To illustrate the vector idea of force, arrows are drawn in the figure. The tail of the vector is shown on the center of the object that feels the force, and the head of the vector points in the direction in which the force acts. Note that the forces of Newton's third law never act on the same object.

---

[8] Newton's third law is applicable to the phenomena we study in this book. There are, however, situations in which the third law cannot be applied. In many problems concerning *electrodynamics*, the third law is not applied in a straightforward manner. It turns out that Newton's third law is linked to the principle of *linear momentum conservation*, and electromagnetic *fields* carry linear momentum. Though we will be discussing linear momentum conservation, the issue of the use of Newton's third law in electrodynamics is beyond the scope of this book.

[9] We refer to magnitudes and other numbers as *scalars*. Scalars do not have direction.

[10] The unit of force in the *International System of Units* (or *SI*, from the French *Système International d'Unités*) is appropriately named the *newton*, abbreviated N. The *pound* is more commonly used in the United States. The conversion is $1\,\mathrm{N} \simeq 0.2248$ pound. In other words, a newton is roughly a quarter-pounder.

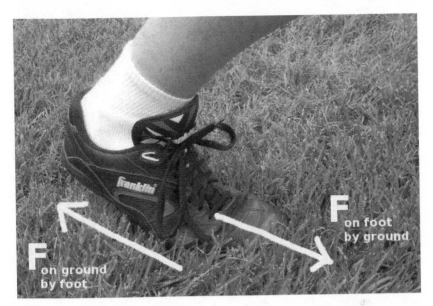

Figure 2.2. My foot while walking. The vector force from the ground on the foot is shown, as well as the vector force from the foot on the ground.

The mechanism behind the force between the foot and the ground in Figure 2.2 is *friction*. When two objects come in contact, a frictional force is present. Sometimes, we try to avoid friction, or at least reduce it as much as possible. Cars are streamlined to reduce air friction, and sleds are made with smooth runners to reduce friction with snow. But friction also helps us in many ways. We could not walk without friction, and race cars could not speed out of the pits without friction. For two solid objects moving against one another, there are two main types of friction to be studied. If two surfaces are in contact but do not slide against each other, *static* friction is present. We think of *kinetic* friction as that acting between two surfaces sliding against one another. In the case of walking, static friction is present as long as the foot does not slide on the ground. If Harmon had kicked in the snow, he would have had to worry about kinetic friction. Wearing cleats helps one stave off kinetic friction, by placing part of the shoe within the packed snow or mud.

Friction is not completely understood, though we do employ some good models. Even objects that seem smooth to us can be quite rough on a microscopic scale. Atoms and molecules form peaks and valleys on an object's surface. Bonds are thus formed between adjacent molecules when two surfaces are brought together. As two objects slide against each other, bonds are continually broken and remade. The bonds made during sliding are not as "tight" as those made when the surfaces are at rest with respect to each other, thus explaining qualitatively why static friction is usually greater in magnitude than kinetic friction.

Harmon is thus able to approach the football with the aid of friction. When he reaches the ball, he kicks it. But what, exactly, is it that propels the ball off the tee? You know the foot is responsible for sending the ball into its flight, but the game would be a little different if the ball left the kicker's foot at the speed of the foot. The ball actually leaves the kicker's foot at a speed greater than the speed of the foot at the time of impact. As the foot makes contact with the ball, the leather and interior rubber bladder of the ball are stretched inward. If you have ever jumped on a trampoline, you know that the stretched material wants to unstretch. But there's more. The air molecules in the ball were initially (before the kick) at a pressure of about 13 pounds per square inch (psi).[11] Compressing the air increases the pressure. Just like a spring, the surface of the football and the air inside want to "spring" back out. In fact, the ball is almost back to its original shape by the time it leaves a kicker's foot. Figure 2.3 shows a sketch of a football just before it loses contact with a kicker's foot. Note that Newton's third law tell us that not only does the ball feel a force from the foot, but the foot feels a force (same magnitude, but opposite direction) from the ball.

At this point, Harmon has approached and kicked the football. As already noted, Stanford's strategy was to squib kick the ball to Cal. Moen fielded the kickoff, and Cal's 4 precious seconds began to tick away.

---

[11] Pressure is a *scalar* and measures the force exerted on a surface per unit area. Technically speaking, the football pressure I gave you is the *gauge pressure*, which is the amount of pressure above the normal atmospheric pressure we all experience. One "atmosphere" of pressure, or 1 atm, is equivalent to $101,325\,\text{N/m}^2$ (SI) or about 14.7 psi. The total internal pressure of the air in the football is thus nearly 28 psi. By the way, you need not worry about being crushed by atmospheric pressure, because cells in your body maintain internal pressures very close to 1 atm. The rigidity of tires, footballs, balloons, and our cells allows internal pressures to exceed atmospheric pressure. Incidentally, if you wish to experience a doubling of the atmospheric pressure, like that in the air in a football, you could dive underwater to a depth of about 10 m (33 ft), though I do not recommend this for novice swimmers!

Figure 2.3. A sketch of a football getting kicked (*left*) and just leaving the kicker's foot (*right*). Note that the football is mostly back to its normal shape.

Average Speed Versus Instantaneous Speed

After watching "The Play" about a dozen times, I found that its novelty began to wear off, and I wondered if it could help me introduce a physics concept in one of my classes. Then I remembered a rather embarrassing moment from my college days. Running a little late for school, I pulled onto a road and never noticed that a cop was just a couple of car lengths behind me. After passing through a stoplight, I hit the gas and was pulled over for speeding shortly thereafter. The police officer told me that he had seen me pull out 2 minutes ago and then informed me that I had been going 50 mph in a 30 mph zone. Emboldened by my recently acquired knowledge of some physics, I said, "I agree that it has been about 2 minutes since I pulled onto the road. However, that was only a mile back. Isn't that right?" He agreed, and equation (1.1) popped into my head. The numbers were on my side, and I said, "If I have gone 1 mile in 2 minutes, doesn't that mean my speed would have been a half mile per minute, or 30 mph?" He confessed that those numbers seemed right, and that a mistake must have been made. I had actually talked myself out of a ticket.[12]

---

[12] The respectable police officer can be forgiven for his lack of knowledge of physics; the impudent kid cannot.

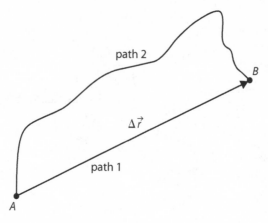

Figure 2.4. Along path 1, you drive a car from point A to point B. The displacement vector, $\Delta \vec{r}$, is shown alongside path 1. Along path 2, you drive a dirt bike from point A to point B.

The fact that I had calculated my *average speed* saved me from a speeding ticket. The policeman had correctly determined my *instantaneous speed*. Those two concepts, however, are as different as a Bear and a Cardinal. Imagine driving along a straight road. You pass a point (call it A) and note your time (call it t_A). Some time later (call it t_B), you pass another point (call it B). Figure 2.4 shows a sketch of your motion, along path 1. Also shown in that figure is your displacement vector, $\Delta \vec{r}$, where the Greek capital delta, Δ, means "change in" and $\vec{r}$ is the mathematical notation I use for a position vector.

Now instead of going from point A to point B in a car, hop on a dirt bike and go off-road. Path 2 in Figure 2.4 is an example of a path you might take. Suppose further that you go pretty fast on your dirt bike, and you go from point A to point B in the same time interval, $\Delta t = t_B - t_A$, that it took your car to traverse path 1. Despite the fact that your dirt bike's speedometer showed a greater speed than the car's, the motions for path 1 and path 2 have the *same* average velocity. For in physics, we define average velocity as

$$\vec{v}_{\text{ave}} = \frac{\Delta \vec{r}}{\Delta t}. \tag{2.1}$$

Notice that average velocity is a vector. That means $\vec{v}_{\text{ave}}$ has both magnitude and direction. If in Figure 2.4 the vector $\Delta \vec{r}$ is 10 miles long and points to the northeast, and if $\Delta t = 10$ minutes, then equation (2.1) gives us for $\vec{v}_{\text{ave}}$ a magnitude of 60 mph and a direction of northeast. Note that when a vector is specified, *both* its magnitude and its direction must be given.

In physics, we make a distinction between velocity and speed.[13] Velocity is a vector and therefore possesses magnitude and direction. Speed, on the other hand, is a scalar; i.e., speed has no direction. We use the term *speed* to refer to the magnitude of the velocity vector. A simple way to think about speed is to note the reading on a speedometer. It gives a number like 60 mph, but you know your direction only by looking out the window. To keep from having to think too much while looking out the window, several makes of cars now come equipped with GPS (Global Positioning System), thus giving direction information and allowing velocity to be known.

The reason the two paths in Figure 2.4 have the same average velocity is that they have the same displacement for the same time interval. In other words, both started at *A* and both ended up at *B* after the same amount of time. If the car's speed were constant all along path 1, the car's speedometer would have read the average speed the entire time. Because the dirt bike traveled a greater distance than the car, i.e. the dirt bike's odometer turned over a greater number than the car's odometer, the dirt bike's speedometer would have to have shown speeds greater than the average speed.

Another way to think about average velocity is to imagine replacing some circuitous path with a nice, straight-line path. In other words, moving at a constant speed given by the average speed along path 1 "replaces" path 2. Figure 2.5 shows a diagram of "The Play" once again. Note that in addition to a sketch of the actual, complicated path of the football after Kevin Moen fielded it, the *displacement vector* is also shown. By my estimation, the ball was displaced about 60 yards from where Moen first caught it. Also shown in Figure 2.5 are the rectilinear "components" of the displacement vector. In getting to the goal line, the ball was displaced about 23 yards perpendicular to the sideline and around 55 yards parallel to it. The relationship between the two components of the displacement vector and the magnitude of the displacement

---

[13] Although words like *speed, velocity, energy, power, momentum,* and so forth have various meanings in common vernacular, they have specific meanings in physics.

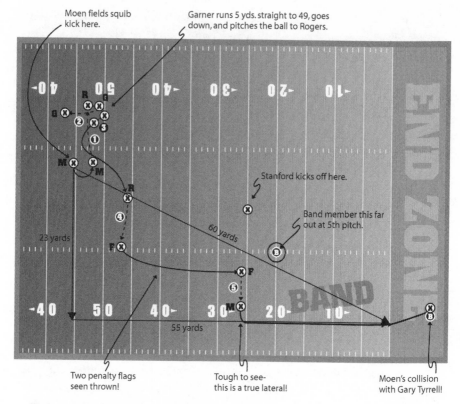

Figure 2.5. "The Play" with displacement vector added (diagonal marked 60 yards). Shown also are the components of the vector parallel and perpendicular to the sideline.

vector is expressed tidily using the Pythagorean theorem, known to many from high school geometry class. To two significant digits,

$$\sqrt{(23 \text{ yards})^2 + (55 \text{ yards})^2} \simeq 60 \text{ yards}. \tag{2.2}$$

I estimate that about 19 seconds elapsed between the time Moen fielded the ball at the Cal 45-yard line and the time Moen (yes, Moen) scored the winning touchdown. Taking the magnitude of equation (2.1), the average speed of the ball was

$$v_{\text{ave}} \simeq \frac{60 \text{ yards}}{19 \text{ s}} \simeq 9.5 \, \frac{\text{ft}}{\text{s}} \simeq 6.5 \, \text{mph} \simeq 2.9 \, \frac{\text{m}}{\text{s}}, \tag{2.3}$$

where I found the speed in various units. The idea of average velocity is that a player could have walked a straight-line path from the point where Moen fielded the ball to the point where Moen crossed the goal line. If that player walked that straight-line path in 19 seconds, and somehow managed to avoid getting tackled, that player would have collided with Moen as he crossed the goal line.

Incidentally, the components of the average velocity vector work exactly the same way. Consider a player walking along the sideline, beginning at the Cal 45-yard line. That player would need to walk 55 yards in 19 seconds in order to reach the goal line at the same time Moen crossed it. In other words, the component of the average velocity vector parallel to the sideline was

$$v_{ave}^{\parallel} \simeq \frac{55\,\text{yards}}{19\,\text{s}} \simeq 8.7\,\frac{\text{ft}}{\text{s}} \simeq 5.9\,\text{mph} \simeq 2.6\,\frac{\text{m}}{\text{s}}. \tag{2.4}$$

I will let you convince yourself that the component of the average velocity vector perpendicular to the sideline was about 2.5 mph. You can then do a little Pythagorean magic with the components of the average velocity vector and see if you get what was found in equation (2.3).[14]

Laterals

Whenever a lateral occurs in a football play, excitement is sure to follow. "The Play" contained five laterals, and each one heightened the excitement level. Laterals are legal as long as the laterals themselves never deliver the football forward. Such a throw would be called a "forward pass" in that case, and multiple forward passes, or any forward pass from beyond the line of scrimmage, are not allowed. Vectors can help us understand laterals quite well. The key idea that vectors help illustrate here is that a lateraled football must not travel in the forward direction *as seen by a stationary observer on the ground* (i.e., you in the stands). Consider the fourth lateral

---

[14] If you were to do the calculation with the numbers given, you would get $\sqrt{(5.9\,\text{mph})^2 + (2.5\,\text{mph})^2} \simeq 6.4\,\text{mph}$. The slight difference between that number and the one given in equation (2.3) is due to *rounding*. I kept only two significant digits in my results. If you wish to use those results for further calculation, you should use an extra digit in intermediate calculations.

in "The Play." Richard Rogers, while running, lateraled the ball to Mariet Ford. Rogers' pitch occurred while he was crossing the Stanford 45-yard line. Ford caught the ball on the run at the Stanford 47-yard line. Had he caught the ball anywhere on the Stanford goal line side of the 45-yard line, Rogers' lateral would have been a forward pass. The difficulty in executing a successful lateral is that a player can actually toss the ball backwards *relative to himself* while the ball still goes forward *relative to the ground*. How can this be?

A few vectors help us understand what is going on. Figure 2.6 shows the velocity vectors needed. The player is running from left to right in the figure; his velocity with respect to (use "wrt" for "with respect to") the ground is $\vec{v}_{\text{player wrt ground}}$. He pitches the ball with a velocity $\vec{v}_{\text{ball wrt player}}$ as he sees it in his reference frame. Unfortunately for him, the frame of reference that matters is the stationary frame of the ground. The referee, or someone in the stands, would see the ball with velocity $\vec{v}_{\text{ball wrt ground}}$, meaning the ball had a component of its velocity in the forward direction. The lateral would be called a forward pass because it moved in the forward direction during the pitch. What a player needs to do, and what Rogers successfully did during the fourth lateral of "The Play," is ensure that the ball's velocity with respect to the ground has a component that is in the backward direction.

As I have mentioned, it is not clear to me if the fifth lateral of "The Play" satisfied the aforementioned legal requirement. The action happened so quickly. Besides, the referees make judgments on laterals while they themselves run to keep up with the players. The ball's velocity they

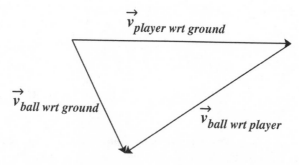

Figure 2.6. Vectors needed to study a lateral. In vector parlance, $\vec{v}_{\text{player wrt ground}} + \vec{v}_{\text{ball wrt player}} = \vec{v}_{\text{ball wrt ground}}$.

see is in their own running reference frame. What the refs must do is keep an eye on both the pitch point and the catch point with respect to the stationary field. It is easier to note those positions on the field than it is to try some mental vector addition while running down the field. Imagine an experiment where a van is traveling at 30 mph. The van's back doors are open. I stand in the back of the van and throw a football out the back at a speed that I know to be 30 mph in my reference frame. You are standing on the side of the road and the van passes you just as I throw the ball out the back. While I might believe that I threw the ball such that its velocity was "backward," you will see the ball appear out the back of the van and fall straight to the ground (air resistance will surely complicate this silly experiment a little). Both of our claims are correct, because our respective claims are made in different reference frames. Vectors are a valuable tool that physicists use to keep straight all of the various reference frames.

We will need to use vectors again as we now move two years ahead, to the day when David slayed the mighty Goliath with one famous sling of a football on a magical night in Miami.

3

All Hail Flutie

Gravity and Projectiles

"Flutie did it!"

Imagine a world without sports updates on cell phones, of not being able to watch a bevy of college football games on a given fall weekend, but only two or three. It is not the Dark Ages, but 1984. By November of that year, Ronald Reagan had won reelection by a landslide; the United States had already dominated the Summer Olympics in Los Angeles, where the Soviet bloc countries were no-shows; the Detroit Tigers ruled baseball; a small university outside Philadelphia was beginning a college basketball season that would shock the sports world; a former Tar Heel named Michael Jordan was set to begin his first NBA season. Today, we are used to flipping on a television, cell phone, or the Internet to get a score or watch a game. Satellite dishes make it possible for die-hard fans to watch every single game their favorite team plays. In 1984, college football fans had but a handful of games to enjoy. Sure, we might be able to watch a local team on a local channel, but the big games, even those of national interest, had to fight for air time. We were used to "Game of the Week" events, and thanks to a brilliant rescheduling move, CBS gave us a gem on Friday, November 23, 1984. It was the day after Thanksgiving, and sports fans had time to sit back and enjoy what was to become one of the greatest college football games of all time.

What made that November game back in 1984 so special? It pitted Boston College against the University of Miami, two colorful and accomplished teams. The Miami Hurricanes were the defending national champions, and the game against the BC Eagles was to be played at the Orange

Bowl in Miami.[1] Miami's quarterback, Bernie Kosar, was everything a pro scout liked to see in a college quarterback. He was a big, tough sophomore who at 6'5" could survey the entire field. Kosar's counterpart at Boston College, Doug Flutie, hardly looked the part of a future pro. At 5'9", Flutie stood in sharp contrast to Kosar. It was to be David versus Goliath.

Boston College got off to a great start, at 14–0, Flutie putting up a string of eleven straight completions. Following the script of a heavyweight fight, Kosar counterpunched with eleven straight completions of his own, and a great college football game began to unfold. The game went back and forth until Miami, trailing by three points, marched 79 yards down the field in the game's final minutes to take a 45–41 lead. Kosar's performance was brilliant. He had passed for 447 yards on the day, and he left Flutie's Eagles just 28 seconds on the clock. Unfortunately for Miami, 28 seconds would prove to be plenty of time for BC's fearless quarterback.

Starting from his own 20-yard line, Flutie completed a pass to Troy Stradford that moved the Eagles to their 39-yard line. Flutie then hit Scott Gieselman for 13 more yards, advancing the ball to the Miami 48-yard line. With only 10 seconds left on the clock, Flutie's next pass was off target, an incompletion that stopped the clock at a mere 6 seconds. There was now only enough time for a final heave into the end zone, often called a "Hail Mary."[2] Such a play, which inevitably yields a mob scene at the goal line, almost never works, but for Boston College that day, "almost never" didn't mean "never."

Flutie took the snap from center and the clock began to tick. The Eagles' quarterback first had to buy enough time for his receivers to make it down the field. He dropped straight back, and when pressure came from Miami's great defensive lineman Jerome Brown, Flutie scrambled off to his right. With but a second on the clock, Flutie launched the ball with all his might

---

[1] Miami had won its first college football national title over Nebraska on January 2, 1984, also at the Orange Bowl. That game too was famous. Nebraska's coach, Tom Osborne, opted against a tie, and at least a guaranteed share of the title, by going for a two-point conversion late in the game with his team down by just one point. The play failed, and Miami went on to win, 31–30.

[2] Many believe this term's use in football originated in 1975 with Roger Staubach, quarterback of the Dallas Cowboys. Late in a playoff game against the Minnesota Vikings, Staubach completed a long pass to receiver Drew Pearson. The play won the game for Dallas. During an interview after the game, Staubach said he shut his eyes, threw the ball as hard as he could, and uttered the Hail Mary prayer.

from the Boston College 37-yard line. As the ball left his hand, one second appeared on the clock. An almost perfect spiral made its way through that gusty night. Three Miami defenders were almost to the goal line when Flutie's pass sailed right over their heads. What those three Miami defenders evidently did not realize was that Boston College's Gerard Phelan was close behind them, at the goal line, unknowingly about to make history. After clearing the Miami defenders, the ball fell into Phelan's arms and Boston College had the upset win over Miami, 47–45.[3] Phelan's famous reception was his eleventh of the day; 226 of Flutie's 472 yards belonged to Phelan. Flutie and the rest of the Eagles began jumping for joy while the Hurricanes stood stunned. Kosar, who had played so brilliantly, walked off the field with his head down. David slayed Goliath with the slingshot arm of Doug Flutie.

By the end of that season, Flutie had become college football's first quarterback to eclipse 10,000 career passing yards, and easily won the 1984 Heisman trophy.[4] Flutie went on to play professional football for two decades with various teams in Canada and in the National Football League (NFL). Bernie Kosar took his famed sidearm delivery to the NFL and enjoyed the better part of 13 seasons, most notably with the Cleveland Browns. Flutie and Kosar are sure to be inextricably linked for a long time to come because of their participation in one of college football's most famous games.

Physics Time

One great thing about using sports to explain physics principles is that there are often so many physics concepts at work in a single play. Doug Flutie's Hail Mary offers a plethora of such concepts. With 6 seconds on the clock, Flutie knows that when he receives the ball from the center, he has just one shot at victory. He and his linemen know that Flutie must

---

[3] As in the Stanford-Cal game, there was no need to kick the extra point.

[4] Flutie's 2240 points far surpassed the 1251 points earned by Ohio State's Keith Byars. It was thought at the time that Flutie's famous "Hail Mary" had sealed the deal on the Heisman trophy, but the votes had already been cast before the game with Miami took place.

give the receivers enough time to sprint down the field and into the end zone. How much time was needed?

A great time in the 40-yard dash is held to be 4.4 seconds. That time was established before electronic timers were used. Fans of football, and die-hard fans of the NFL Combine results, sometimes hear 40-yard dash times that are less than 4.2 seconds. Some players opt for "individual workouts," where their 40-yard dash times are determined by hand timers. They know that a timer's reaction time at the start of the dash will yield faster elapsed times than those produced by electronic timing. Breakdowns of Ben Johnson's 1988 steroid-assisted record-breaking 100-meter sprint have shown that he reached 40 yards in 4.38 seconds.[5] We can therefore safely assume that Boston College's wide receivers, who after all were wearing perhaps 15 pounds of football gear, did not cover 40 yards in less than about 5 seconds. They were also said to be running into a headwind that some estimates put at 30 mph (at field level, the wind speed would not have been nearly that great). If we wish to come up with some plausible numbers for a football player running 40 yards, we need to discuss some physics.

Assume first that Gerard Phelan's intent was to run straight down the field in the shortest possible time. He starts from rest and takes off at the snap of the ball. He will accelerate for a brief time before hitting a fairly constant speed. Let us suppose that Phelan accelerates at a constant rate, and then maintains a constant speed. If his sprint takes place in a straight line, we can use what are called the constant acceleration kinematics equations. That fancy term simply means that the equations describing the position and velocity of an object are valid only for one constant value of the acceleration. If, after a time t, x is an object's position and v is its velocity, the following equations describe one-dimensional motion:

$$x = x_0 + v_0 t + \tfrac{1}{2}at^2, \qquad (3.1a)$$

$$v = v_0 + at, \qquad (3.1b)$$

[5] An excellent article dealing with absurd times in the 40-yard dash is Mark Zeigler's "The NFL treats 40-yard dash times as sacred," which appeared in the April 20, 2005, edition of *The San Diego Union-Tribune*. Zeigler discusses Johnson's 40-yard time, among others.

$$v^2 - v_0^2 = 2a\,(x - x_0), \qquad (3.1\text{c})$$

$$x = x_0 + \tfrac{1}{2}\,(v_0 + v)t, \qquad (3.1\text{d})$$

where x_0 and v_0 are the initial position and velocity, respectively, and a is the constant acceleration. The above equations are derived in any high school or college physics course. Like many equations in physics, they originate from a simple starting point.[6] We will now have Phelan running a 40-yard dash that consists of two separate constant-acceleration pieces: he starts from rest and reaches a top speed $(a \neq 0)$, then runs at top speed $(a = 0)$.

Let's tackle this problem in a general way. This approach is powerful in physics because we reach conclusions that work for any choices of numbers. In other words, we need do only one derivation instead of working through a new one each time we wish to change the numbers. Say Phelan has nonzero acceleration for a time t and runs a distance d. Starting from rest $(v_0 = 0)$ at the origin $(x_0 = 0)$, equation (3.1a) gives

$$d = \tfrac{1}{2}at^2. \qquad (3.2)$$

The velocity Phelan reaches at the end of the nonzero acceleration part is found by using equation (3.1b), i.e.

$$v = at. \qquad (3.3)$$

Finally, equation (3.1a) can be used during the zero acceleration portion of Phelan's run.[7] Suppose the total distance traveled is called D, and T represents the total time running. Then,

$$D - d = v\,(T - t). \qquad (3.4)$$

---

[6] To derive the equations, begin with $a = $ constant and two definitions: $a = dv/dt$ and $v = dx/dt$, where d/dt is the first derivative with respect to time (a calculus tool). Integrate $a = dv/dt$ once to get equation (3.1b), then use $v = dx/dt$ and integrate one more time to get equation (3.1a). Solve for t in equation (3.1b), substitute into equation (3.1a), and you will get equation (3.1c). Equation (3.1d) is derived after noting that for constant acceleration, $(v_0 + v)/2$ represents the *average* velocity during a time t. If you have not studied calculus, and you still want to know how I got equations (3.1), drop me an e-mail and I will offer some more guidance.

[7] Behold that setting $a = 0$ in equation (3.1a) takes us back to equation (1.1), which we used in *The Pre-Game Show*. Every time we wish to model the world in a more complicated way, the equations we need to employ will surely get more complicated as well.

Equations (3.2), (3.3), and (3.4) can be solved for the three unknowns (t, v, and a). I'll spare you the couple of lines of algebra needed and give you the following:

$$t = \frac{2\,dT}{d+D}, \tag{3.5a}$$

$$v = \frac{d+D}{T}, \tag{3.5b}$$

$$a = \frac{(d+D)^2}{2dT^2}. \tag{3.5c}$$

Enough math already! Let's get to some numbers relevant for football. Suppose Phelan needs 10 yards to reach his maximum speed (perhaps a little short on the distance he would need, but we can always change our numbers). His 40-yard dash then takes place in 5 seconds, meaning $D = 40$ yards, $T = 5$ seconds, and $d = 10$ yards. Inserting these numbers into equations (3.5) gives $t = 2$ s, $v = 10$ yards/s (30 ft/s or roughly 20.5 mph), and $a = 5$ yards/s^2 (15 ft/s^2). Feel free to put any combination of D, T, and d into equations (3.5) that you like—there is no more deriving to do.

If equations are not your forte, try something more visual. Figure 3.1 shows a graph of Phelan's 40-yard dash position versus time and a graph of his velocity versus time. Real-world data are never as clean as this. What the figure illustrates is the result of our attempt to model Phelan's 40-yard dash. We have not accounted for acceleration changes due to his legs moving back and forth, or for the other aspects of running that contribute to his acceleration not being exactly constant. Our graphs simply show us the basic idea of running down the field once the football is snapped. As Phelan accelerates, his graph of position versus time is curved,[8] a sure sign of acceleration.[9] At the cutoff time of 2 seconds, Phelan reaches maximum velocity, and the position versus time graph is then a straight line. That is the result of setting $a = 0$ in equation (3.1a).

A velocity versus time graph, like the lower one shown in Figure 3.1, illustrates accelerated motion wherever the curve is not horizontal. After

---

[8] Equation (3.1a) is the equation of a *parabola*.
[9] From a calculus viewpoint, the acceleration is the *curvature* of an x vs. t plot, since $a = d^2x/dt^2$.

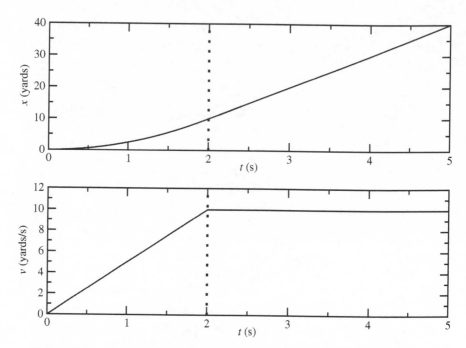

Figure 3.1. The position versus time for Phelan's 40-yard dash is shown at top; the velocity versus time is shown at bottom. It is assumed that Phelan's acceleration goes to zero after he has run 10 yards. The vertical dotted lines indicate when the acceleration is turned off.

2 seconds, Phelan's v versus t plot goes flat; there is no more acceleration. Note that each v value at a given t value is the slope of the x versus t plot *at the same value of time*.[10] The acceleration is the slope of the v versus t plot.[11] For example, up until 2 seconds, the slope of the v versus t plot is $(10 \text{ yards/s})/(2 \text{ s}) = 5 \text{ yards/s}^2$, the acceleration we already found. During the remaining 3 seconds, the slope of the v versus t plot is zero, as it should be for zero acceleration. Calculus is the mathematical tool that allows us to move quickly from one graph to another.[12]

---

[10] If you have studied calculus, you know this right away from the definition $v = dx/dt$.

[11] Again, the definition $a = dv/dt$ succinctly describes this idea.

[12] The slope, or *derivative*, is not all that calculus can do for us. The area under the v versus t curve, or the *integral*, is the *displacement* of our runner. Convince yourself that the area under the v versus t plot in Figure 3.1 is indeed 40 yards.

So, Flutie's receivers could cover about 40 yards in 5 seconds, provided they were not "chucked" at the line of scrimmage. They then need to slow down and position themselves in the end zone. The 6 seconds left on the clock truly allowed only one game-deciding play. Indeed, Flutie uncorked the ball just before the play clock hit all zeros. Given the roughly 3 seconds that the ball was in the air, Flutie's receivers had enough time to get to the end zone and set up their "Flood Tip," which is what they called their practice drill whereby they attempted to tip the ball to a teammate after flooding into the end zone.

How Goliath Got Slain

Well, Flutie's receivers have done their part. They accelerated up to some top speed, made it into the end zone, and got themselves into position for the miracle finish. All Flutie had to do was heave the ball a whopping 62 yards into a headwind. To this day, I still can't believe that he was able to throw the ball that far in those conditions. But, hey, some people can rise up and slay the giant.

To understand what Flutie's thrown football was doing during its flight to immortality, we have to know something about projectile motion. As I watch Flutie's famous pass, I can measure certain quantities of interest. For example, the horizontal range of the pass was about 62 yards. After timing the pass on each of several viewings, I estimate that the ball was in the air about 3 seconds. There are many other quantities that I cannot measure so well; I would, however, really like to know what they are. What was the launch speed of Flutie's pass? How high did it go? How much energy did it have when Flutie let go of it? How much power did Flutie need to exert to launch the ball at the necessary speed? All of these questions (and a few more!) pop into my head when I watch that play. Luckily, physics helps answer them.

Let's begin with what would happen to a thrown football if there were no air around. Granted, that's not what Flutie faced; plenty of air, wind, rain, and other factors affected the flight of the ball. So we need to build up our understanding of what happened in layers. Each new layer adds complication; the path to describing reality gets more and more tricky.

So, suck all of the air out of the Orange Bowl and give the players oxygen tanks and masks.

In a vacuum, the only force the football feels once it leaves Flutie's hand, before it reaches Phelan's cradling arms, is the force of gravity from the Earth. Newton's first law tells us that an object in motion with a constant velocity (could be zero velocity) remains at that velocity unless acted upon by a net external force. What all that means is that if there were no gravity around when Flutie let go of the ball, it would continue along its straight-line path and never come down. Of course, the players on the field would not feel gravity either, meaning that the "deep-space football" game would look mighty strange. With a gravitational pull on the football straight downward, the ball must accelerate straight downward. Newton's second law tells us that the product of an object's mass with its acceleration is equal to the net external force on it. One complication with Newton's second law is that it is a *vector* equation. In symbols, I write the second law as

$$m\,\vec{a} = \vec{F}^{\text{net}}, \tag{3.6}$$

where m is the object's mass, $\vec{a}$ is the acceleration (a vector), and $\vec{F}^{\text{net}}$ is the net external force on the object (again, a vector). If gravity is the only force on the football, which is our beginning assumption, the acceleration vector points straight down and has magnitude of what we in physics often call g. That symbol saves us some writing as we play with equations. The value of g is roughly $32\,\text{ft/s}^2$ ($9.8\,\text{m/s}^2$ or $22\,\text{mph/s}$). Think about that for a moment. Equation (3.2) tells you that if you drop a football from a height of about 16 feet (4.9 m), it will take about 1 second to hit the ground. Dropped from rest, the football will hit the ground with a speed of about $32\,\text{ft/s}$ ($9.8\,\text{m/s}$ or $22\,\text{mph}$). At the moment it hits the ground, the football will be speeding in a school zone!

So, in our air-free model of the football in flight, $\vec{a}$ in equation (3.6) has magnitude g. To obtain the size of the force on the football, we need its mass, m. According to the rules,[13] a football has a weight between 14 and 15 ounces. Note that the *weight* is specified, not the *mass*. Weight is a

---

[13] See Section 3, Article 1, item h of the *2007 NCAA Football Rules and Interpretations*. The same weight is given in Rule 2, Section 1, paragraph 3 of *2006 Official Playing Rules of the National Football League*. As far as I can tell, the football's weight has been standard since 1912.

force, meaning it has direction, whereas mass is a scalar. Put pedantically, the weight of a football is the size of the gravitational force on the football, and the direction of the weight is toward the center of the Earth. Specifying "14 or 15 ounces" is really specifying the magnitude of the weight vector. (It's good they don't let physicists write the rules of football.) Anyway, the weight of an object is its mass multiplied by the local acceleration due to gravity. In other words, if you go out into deep space, far from any massive objects, your weight is zero, but your mass has not changed. Thus, you are "weightless," but not "massless." Mass is an intrinsic property of an object, and it exists whether the Earth is here or not. Weight is the force we feel from the Earth, and its magnitude is mg. Thus, $mg = 14$ or 15 ounces. In SI units, the weight of a football is about 3.9–4.2 newtons. The mass is found by dividing the weight by g. Thus, the football's mass is about 0.027–0.029 slugs[14] in British units and 0.40–0.43 kilograms in SI units.

What equation (3.6) tells us is that if the net external force on the football is mg, something we put on the righthand side of that equation, then the mass cancels and $a = g$. So, after all that discussion about mass and weight, the mass cancels! The point here is that in a vacuum, *all* objects fall at the same rate, independent of their masses. But whereas the rate of fall is the same for all objects, the acceleration is *not* strictly constant. The gravitational force on a football decreases as the football gets farther from the surface of the Earth. If the football, however, managed to go from sea level to a mile off the ground (pay attention, Denver fans), its weight decreases by only about 0.05%. We are thus safe in assuming that the gravitational acceleration in such circumstances is constant.

Taking the acceleration to be constant, we can make use of equations (3.1). But, you ask, "How can we do that if those equations are

[14] The "slug" is almost never used anymore. It is the mass of an object that accelerates 1 ft/s^2 while experiencing 1 pound of net external force. Refer to equation (3.6) to see how the units work. The intriguing thing about U.S. practice is that we use the British unit of force (pound) and the SI unit of mass (kilogram). Note, however, that conversions you see between pounds and kilograms, like $1 \text{ kg} \simeq 2.2 \text{ lbs}$, are not really correct in the strictest sense. Mass and force are simply two different things. The conversion is useful; remember, however, that the "real" conversion is $1 \text{ slug} \simeq 14.6 \text{ kg}$. The conversion you often see comes from equating *weights*. In other words, $(1 \text{ kg})(9.8 \text{ m/s}^2) = 9.8 \text{ N} \simeq 2.2 \text{ lbs}$. Thus, 2.2 lbs is the *weight* of 1 kg *on the surface of the Earth* (i.e., at sea level). Sorry to complicate what you once thought was a simple conversion!

for *one*-dimensional motion? Doesn't our football move in *two dimensions?*" You are right in observing that our projected football moves in a two-dimensional plane.[15] Luckily, we have already discussed vectors, meaning that we can split equation (3.6) up into various components. Now, the real world says that the acceleration due to gravity points *down*. The real world couldn't care less what some nerdy physicist chooses for a coordinate system. Thus, I will model the real world by having one axis point vertically and one axis point horizontally. Often, the vertical axis is called y and the horizontal axis is called x.[16] By choosing one axis to be vertical, all the acceleration is along that axis. The motion in the horizontal direction is therefore easy, because there is no acceleration component in that direction; the x motion is described simply by using equation (3.1a) with the acceleration set to zero, meaning

$$x = x_0 + v_{0x}t, \tag{3.7}$$

where v_{0x} is the x-component of the initial velocity vector. Since there is no x-component of the acceleration, the x-component of the velocity does not change. Along the y direction, equations (3.1) can be written as

$$y = y_0 + v_{0y}\,t - \tfrac{1}{2}gt^2, \tag{3.8a}$$

$$v_y = v_{0y} - gt, \tag{3.8b}$$

$$v_y^2 - v_{0y}^2 = -2g\,(y - y_0), \tag{3.8c}$$

$$y = y_0 + \tfrac{1}{2}(v_{0y} + v_y)t, \tag{3.8d}$$

where v_{0y} is the y-component of the initial velocity vector and the acceleration in the vertical direction is $-g$, negative since I define *up* as the direction of increasing y.

---

[15] Including the air and the effects due to the *spin* on the ball means that we really need to consider *three*-dimensional motion. We will have more to say about that in chapter 7, when we discuss soccer.

[16] Just as with the choice of coordinate system, the real world couldn't care less what we call our axes. But so many science and math books use an "$x - y$ coordinate system" that some people have ingrained in their heads the notion that a horizontal axis is x and a vertical axis is y. I have attended numerous talks where the speaker will say something like, if referring to my Figure 3.1, "Time is shown on the x axis." To hammer in nature's apathy for our choice of coordinate labels, I sometimes use the "elephant-hippopotamus coordinate system" in my classes.

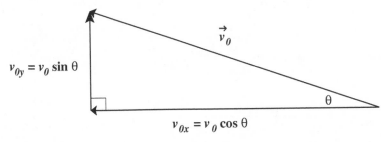

Figure 3.2. The initial velocity vector $\vec{v}_0$ is split into its x and y components. They are $v_{0x} = v_0 \cos\theta$ and $v_{0y} = v_0 \sin\theta$, where θ is the initial launch angle of Flutie's Hail Mary pass.

Okay, you have seen a bunch of equations; they all originate, however, from the fact that we are assuming $\vec{a}$ to be constant. Let's assume that Flutie let go of the football at the origin of our coordinate system. That means that $x_0 = y_0 = 0$. Further, assume that Phelan caught the ball at exactly the same height above ground at which Flutie released the ball. From the video of the play, we can conclude that this is a reasonable approximation.

The only thing left to do before calculating anything is to split the initial velocity vector into its x and y components. Figure 3.2 shows how this is done, using a little trigonometry. I label the initial launch angle by the lower-case Greek letter theta (θ). (Physicists love using Greek letters for angles.) Once the initial velocity vector is split into components, we go off and work with each direction separately. What links the two directions in our work is *time*.

There are many things we can now calculate concerning Flutie's pass. I will spare you the algebra and show you the results of playing with the equations we have seen so far. The time of flight (T), horizontal range (R), and maximum height (H) of Flutie's pass are

$$T = \frac{2 v_0 \sin\theta}{g}, \tag{3.9}$$

$$R = \frac{v_0^2 \sin 2\theta}{g}, \tag{3.10}$$

and

$$H = \frac{v_0^2 \sin^2\theta}{2g}, \tag{3.11}$$

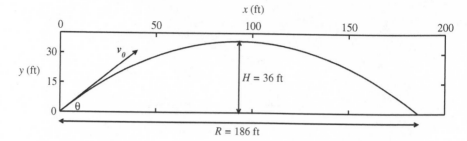

Figure 3.3. The vacuum model prediction of Flutie's pass. The launch angle is $\theta \simeq 38°$; the launch speed is $v_0 \simeq 78$ ft/s; the maximum height is $H = 36$ ft. Note that the aspect ratio in the figure is 1. Note also that the x axis labels are placed on top of the plot, so that the range can be identified easily.

respectively.[17] As I have already mentioned, I estimate $T \simeq 3$ s and $R \simeq 62$ yards. I can then plug those values for T and R into equations (3.9) and (3.10) and solve for v_0 and θ. I get[18] $v_0 \simeq 78$ ft/s (53 mph or 24 m/s) and $\theta \simeq 38°$. Plugging these results into equation (3.11) gives $H = 36$ ft.[19]

I'll give you one more equation. If the time is eliminated from the projectile equations, the following can be derived:

$$y = x \tan \theta - \frac{g\,x^2}{2\,v_0^2 \cos^2 \theta}. \qquad (3.12)$$

This equation is the actual model trajectory of the football if we could trace it out in the x-y plane. In other words, it describes what we would see if we were sitting in the stands while Flutie threw his famous pass (with no air around!). The trajectory, a parabola, is shown in Figure 3.3 for the numbers we have found so far.

The question we need to ask ourselves now is "How reasonable are our results?" Our boldest approximation was ignoring air resistance. It turns

---

[17] If you have ever taken a physics class, or if you have ever perused a physics book, you have probably seen these equations. Be careful about using them. Always make sure you understand the assumptions behind any equations you use. The three equations I give here rest on three main assumptions: (1) air resistance is ignored, (2) the acceleration is constant, and (3) the initial and final heights of our football's trajectory are the same. If one (or more) of those assumptions is (are) not valid, the equations I show here are no longer true.

[18] Convince yourself that $\tan \theta = gT^2/2R$ and $v_0 = \sqrt{(gT/2)^2 + (R/T)^2}$.

[19] Another nice result is $H = gT^2/8$.

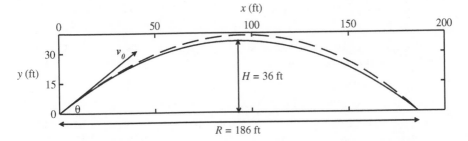

Figure 3.4. Figure 3.3 is shown again, this time with a dashed curve representing the trajectory resulting when air resistance is included in the model. The new launch speed is about 85 ft/s, and the new maximum height is roughly 39 ft, just a yard higher than that in the vacuum case.

out that for a football, a fairly aerodynamic ellipsoid if there ever was one, including air resistance does not alter our results much. A baseball, on the other hand, is greatly affected by air resistance. The baseball's almost spherical shape is less aerodynamic than a football's ellipsoidal shape. A home run in the air could travel almost twice as far if it were in a vacuum. Forget steroids—just suck all the air out of the stadium![20] If Flutie had thrown a wobbly duck instead of a tight spiral, air resistance would be more of a concern. A football with a nice spiral is actually quite aerodynamic. Despite the fact that the initial air drag on the football is about one-quarter of the football's weight, our results would not be altered much.

Including air resistance creates equations that cannot be solved analytically; a computer would have to be used. What I did was put air back into the Orange Bowl and assume that Flutie let go of his pass at the same angle ($\theta \simeq 38°$) we already found with the vacuum model. I then determined what launch velocity he needed to have for the football to travel the same horizontal distance. Figure 3.4 shows the result of the calculation.[21] To make the curve that accommodates air resistance, a curve that is no longer a parabola (despite its similarity to one), Flutie's initial launch

---

[20] For more details about baseball physics, see Robert K. Adair's excellent book entitled *The Physics of Baseball* (third edition, Perennial/Harper Collins, 2002).

[21] We will talk about air drag more in the next chapter. Since air resistance does not drastically alter our work here, I have kept the technical focus on the vacuum model. For those of you itching for more details, I used equation (4.2) with $C_D = 0.1$. That value of the drag coefficient is a reasonable estimate for a nose-first ellipsoid.

speed was about 58 mph (or about 26 m/s). Thus, the inclusion of air resistance means that our calculated launch speed is only a bit more than 9% larger than what we found for the vacuum case. The new maximum height is only about a yard higher.

Now, to finish up the kinematics of Flutie's pass, let's see how much effect a headwind would have had. In my computer calculation, I merely need to work with the *relative velocity* of the ball with respect to the air. Suppose Flutie faced a 5 mph headwind that was blowing horizontally in his face. His new launch speed would need to be about 59 mph to get the football to land in the same location. If the headwind were 10 mph, he would need to release the football at about 60 mph. It seems that although modest headwinds do affect the launch speed Flutie needed to have, they do not significantly alter our predictions.

I thus believe we are able to claim that Flutie threw his indelible pass with a launch speed of about 60 mph. The camera angle of Flutie's pass is such that one cannot get an accurate measure of the launch speed and launch angle. Given what we calculated, however, the football's horizontal component of its launch velocity is about 47 mph.[22] That is about 69 ft/s or 23 yards/s. Without air resistance, the horizontal distance covered in 3 seconds would be about 69 yards. Because air resistance slowed down Flutie's pass during its entire trajectory, our calculations seem reasonable, given that his actual pass covered about 62 yards in 3 seconds.

If we look back over our work here, we find that we did a decent job of predicting Flutie's launch speed by considering a football thrown in a vacuum. Putting air resistance into the model tacked on the need for a little more initial speed, and overcoming a headwind added only a little more. We will worry more about air resistance as we consider our next great sporting feat. We now leave the college gridiron for the beautiful vistas of France.

[22] I am using Figure 3.2 with $v_0 \simeq 60$ mph and $\theta \simeq 38°$.

4

Vive le Lance

Simplifying the Complicated

King of the Mountain

On Saturday, July 3, 2004, in the historic Belgian city of Liège, Lance Armstrong embarked on a journey that ended twenty-two days later in the Champs-Elysées in Paris. His goals, to don the yellow jersey upon entering the French capital and to stand atop the cycling world with his record-breaking sixth straight Tour de France championship, came to fruition.[1]

Was there a turning point for Armstrong and, if so, could it be found among nearly 84 hours of cycling across roughly 3400 km of uneven terrain? I suggest stage 16. Stage 15 finished on July 20th in the French village of Villard-de-Lans, but the cyclists had to be ready to get up the next day and begin the time trial (which is a race against the clock, without influence from other riders) at Bourg d'Oisans. The Pyrenees were behind the cyclists, and Armstrong held an 85-second lead over the Italian rider Ivan Basso. Five tough stages were still to be contested, and Armstrong had won the 2003 Tour de France by a mere 61 seconds. A *lot* could happen in the remaining five days.

Let's look more closely at stage 16 to understand why it might be special. A map reveals that the distance from Bourg d'Oisans to the stage's end at the ski station at L'Alpe d'Huez is only 15.5 km. That may seem like a pleasant little jaunt, but those 15.5 km lie in the heart of the French

---

[1] Armstrong won an unbelievable *seventh* consecutive Tour de France in 2005. He then retired from Tour de France racing, giving someone else a chance to win. He is planning a comeback in 2009.

Figure 4.1. The King of the Mountain, Lance Armstrong. (Used with permission of REUTERS/Eric Gaillard)

Alps. To get from start to finish, the riders needed to increase their elevation by 1.13 km (about 0.7 mile). Besides the steep ascent, the riders had to contend with 21 hairpin turns.

When all was said and done, Lance Armstrong had emerged victorious in the punishing individual time trial. He finished the stage in 39′ 41″, just over a minute faster than his closest rival on that day, the German Jan Ullrich.[2] The famed yellow jersey was secure on Armstrong's back as his overall lead grew to 3′ 48″. The race was over. Sure, there were theatrics

[2] Unfortunately for Tour de France fans, Ullrich, the aforementioned Ivan Basso, and several other cyclists were disqualified from competing in the 2006 Tour de France. Charges of doping marred the entire 2006 race. The race's winner, Floyd Landis of the United States, failed a drug test after his incredibly fast stage 17. Race officials ceased to consider Landis as the 2006 champion and instead recognize the runner-up, Spaniard Óscar Pereiro Sio, as the true champion. Given that Landis was not caught until some time after the race, it seems to me that it would be better if we could purge the 2006 Tour de France from our memories.

the next day when Armstrong sprinted to stage victory in the final 25 m, but he had shown in stage 16 that he was king of the mountain, and nobody was going to knock him off (see Figure 4.1).

Can We Predict a Stage Time?

Let's now put on our physics caps and think about how to *predict* the time it would take Lance Armstrong to complete stage 16 of the 2004 Tour de France. Note that making a prediction is a task different from what we have done with the sporting events we examined in the preceding chapters. Our goal here is not to analyze an event *after* it happens. We want to be able to make a prediction *before* the event happens. Predicting means making a model. I'll try to illustrate how a scientist goes about making such a model by starting simple and moving on from there.

One of the first models we ever learned in school is

$$\text{distance} = \text{speed} \times \text{time}. \tag{4.1}$$

Also, see equation (1.1). If our car is going, say, 60 km/hr and we drive for 10 hours, we will have gone 600 km. This is true only if our speed is *constant* or if our "speed" represents an *average* speed, as we discussed in chapter 2. You may wonder if we can use equation (4.1) to predict how fast Lance Armstrong could bike stage 16. We know the distance and we want the time; we do not, however, know the speed. In fact, equation (4.1) may not be such a good model for determining Armstrong's time. It would be unlikely that Armstrong could bike at a constant speed. It would also be convenient if we could predict the time *without* having to shoot Armstrong with a radar gun all along the route in order to determine his average speed.

A First Try

Scientists love to make "back-of-the-envelope" estimates of real-world problems, so that we have some reasonable feel for the numbers involved. We also want to be sure that if we build a complex model for a real-world problem, the numbers produced by our model are not disturbingly wild. For example, I've already told you that Armstrong finished a 15.5 km (or, if you prefer, about 9.63 miles) stage in 39′ 41″, meaning that his average speed was about 23.4 km/hr (or about 14.6 mph). Of course, I don't want to put the cart before the horse; what has been done in the past, however, can be used to build a physical model.

If, for example, I describe to you some model that we might use to predict a stage-winning time of the Tour de France, and that model predicts a winning time of, say, 17 hours, you probably would not trust that model. Or, suppose I told you that a model had predicted that Armstrong could bicycle at a clip of 200 km/hr. That speed is obviously too great, but we know that that is so only by using our personal experiences and making some initial estimates. Having an idea of what range our numbers should fall in is valuable for building confidence in a model.

So, how fast can an elite athlete bike stage 16 of the 2004 Tour de France? You know the distance, and you know that if you could estimate the average speed, you could find the time it takes to complete the stage. Well, how fast could a person *run* the stage? Elite athletes can run 100 meters in approximately 10 seconds.[3] An average speed of 10 m/s, or 36 km/hr, means that the 15.5 km stage could be completed in about 26 minutes. Of course a world-class sprinter runs the 100-meter dash on flat ground, and you know that the pace set in that 10-second race could not possibly be maintained for much more than the 100-meter race distance. A time of 26 minutes is, however, a reasonable lower limit for our cyclist's stage time.

What about an upper time limit? A world-class cyclist could easily outpace someone walking, even if the race terrain were straight up a mountain. Though I am far from being a top-notch athlete, I usually walk

---

[3] As of the time of this writing, both the world and Olympic records for a 100-m sprint are held by Jamaican Usain Bolt. The "Lightning Bolt" ran 9.69 s on August 16, 2008, at the Beijing Summer Olympics. What's amazing about Bolt's dash to glory is that he was celebrating during the last 10 m of the race, and his left shoelace had come untied during the race. Some scientists claim that Bolt's time could have been closer to 9.55 s had he not ended the race so theatrically (see "Scientists say Usain Bolt could have gone faster," by Scott Gullan in the September 11, 2008, edition of the *Herald Sun*).

about 4 miles in an hour on my treadmill. Let's up that pace by 50% and say that a reasonable lower figure for a stage's average speed is 6 mph, or about 10 km/hr. Thus, an upper time limit for the 15.5 km stage on this basis is roughly 1 hr 33'.

We might also like an estimate that falls between our two extremes. Can you think of a runner who would definitely not sprint the entire time, yet certainly have an average speed better than a fast walker? How about a marathon runner? A good marathon runner can finish the 26.2-mile race in under three hours, meaning that the runner's average speed is at least 8.7 mph, or about 14 km/hr. At an average speed of 14 km/hr, a cyclist finishes a 15.5 km stage in about 1 hr 6'.

We now have a few numbers based on rough estimates for what a top athlete could do in a few different types of walking/running races. The conversion of power in walking and running is of course different from that of cycling. We might at least think, however, that a cyclist going up a steep hill could surely not equal the time we predicted using an Olympic-class sprinter's average speed on flat ground. Thus, 36 km/hr for the 15.5-km stage is probably too fast. We may also believe that the time estimate based on the walker's average speed is a little long. This is, however, only a rough estimate. We have to remember that a cyclist competing in stage 16 of the 2004 Tour de France climbs a steep grade for essentially the entire stage. Although the sprinter's average speed has to be too fast for our model and the walker's is most likely too slow, the marathon runner's may not be such a bad estimate.

Think about what we have done. With essentially no more information about stage 16 of the 2004 Tour de France than the distance from start to finish, we have a range of predicted times. Our guess at this point is that a cyclist could finish the stage in a time between half an hour and an hour and a half. If we now make a more sophisticated model and predict a time outside of our "back-of-the-envelope" range, we are forced to ask ourselves some questions. Are we applying physics correctly? Have we included all of the essential ingredients in our model? Or, did we do something silly in making our rough estimates? The power of making a few estimates before creating a complex model is not only to gain a feel for the sizes of the predicted numbers, but to gain confidence that a complex model might actually describe the real world considerably well.

A More Complex Model

A good place to commence building a more complex model might be to look at a snapshot of stage 16. We have several options. We could look at relief maps from our local library; we could look at satellite images from space; we could look at actual photographs of the route. But before we get too crazy with pictures, remember that physicists reduce the system we are studying to its simplest possible constituents. Only if our simple models fail do we look for something more complicated. Check out Figure 4.2 from the Tour de France web site (www.letour.fr).

If you have ever seen that famous stretch of hairpin-laden road that makes up stage 16, you will know that Figure 4.2 is a gross simplification of the actual terrain. But, hey, let's see what we can do, and if we do not get a reasonable time for Armstrong's trek up the mountain, we will go back to the drawing board.

The first thing to notice about the profile for stage 16 is the distance given at various points. The numbers given in meters that go up the hill are the heights above sea level, and the numbers given in kilometers at the

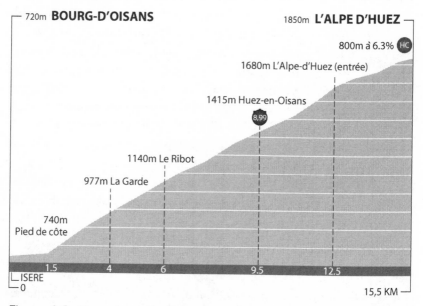

Figure 4.2. Dynamic profile of stage 16 of the 2004 Tour de France.
(From Amaury Sport Organisation; the image is taken from www.letour.fr.)

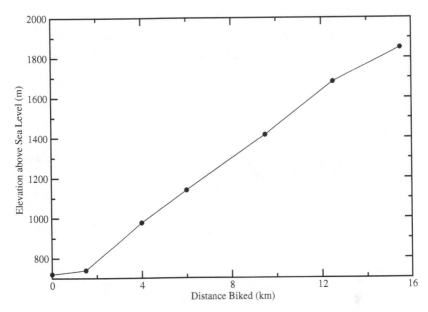

Figure 4.3. Our model of stage 16 of the 2004 Tour de France.

bottom are the distances biked up to that point in the stage. We are thus given a height and a distance biked at several points along the stage. Since we have no other information about the terrain at points between those given, let's draw a straight line between the known points. Figure 4.3 shows you what our terrain model looks like, but I remind you again that we physicists need to be careful when looking at data. Even though I plotted biking distance on the horizontal axis in Figure 4.3, just as in the web site's picture shown in Figure 4.2, that distance says nothing about what is traversed in the *horizontal* direction. For the model we are trying to make, Figure 4.4 shows a far better picture of what we are trying to do. Note that the figure shows the last two data points given for stage 16.

Notice also in Figure 4.4 that I have created a figure we have all seen in high school—a right triangle. You remember geometry class, right? The "distance biked" is the *hypotenuse* of the right triangle, while the "height" is one of the triangle's two legs adjacent to the right angle. The idea is to make a sequence of right triangles using *all* the points given to us on the web site's stage profile. Check out Figure 4.5. Not to beat a dead horse too badly, but note that what you see in Figure 4.5 is slightly different from what you see in Figures 4.2–4.3. The web site profile puts the distances

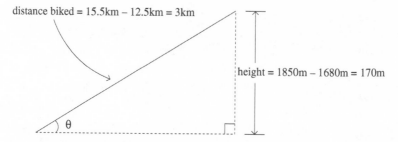

distance biked = 15.5km – 12.5km = 3km

height = 1850m – 1680m = 170m

θ

Figure 4.4. The last segment of stage 16 of the 2004 Tour de France (not to scale). The angle of the incline is θ.

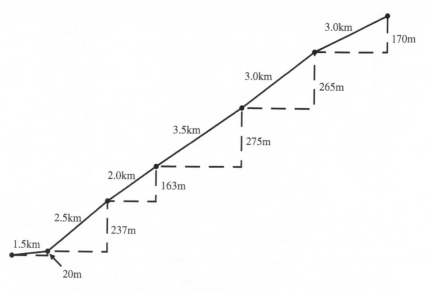

3.0km
170m

3.0km
265m

3.5km
275m

2.0km
163m

2.5km
237m

1.5km
20m

Figure 4.5. The sequence of right triangles used to model stage 16 of the 2004 Tour de France. The "distance biked" is the hypotenuse, and the "heights" are given on the vertical legs adjacent to each triangle's right angle. The sketch is not to scale.

biked on the *horizontal* axis despite the fact that those distances are really the lengths of the *hypotenuses*.

We have taken a simplified cartoon of the terrain and used the distances given to us to create a sequence of right triangles. Figure 4.5 thus more effectively represents our terrain model. I have tried to make the complicated terrain of a Tour de France stage into something with which a physicist is comfortable.[4] The inclined plane is studied in almost all first-semester introductory physics courses in high school and college. Once we know how to solve a physics problem on a *single* inclined plane, we can put a bunch of them together and do the same thing over and over again. If you were to go to the Tour de France web site I mentioned earlier and count up all the inclined planes you could form from the data given for all 21 stages (including the "Prolog"), you would find 498 of them! What I am trying to convince you of here is that if we can learn to solve a physics problem on *one* inclined plane, we can model the *entire* Tour de France. Our terrain model would be even better if more data points were given on the profile images, but we will make use of what is available.

Physics Time

To understand how a bike moves on an inclined plane, we need to construct a model for the bike and for the rider. The simplest thing to do is assume that bike and rider are a conglomerate object of some fixed mass. We will ignore arms, legs, spokes, chain, helmet, water bottle, and all that other stuff that makes up the biker and the bike. Just determine the mass of the biker and the bike together and think of them as one object. We have already illustrated that we have a two-dimensional terrain model. So, put our rider-bike mass at the foot of the first inclined plane in Figure 4.5 and start with zero speed. Our task is to determine the time it takes our rider to make it to the end of the stage.

To get the bike moving from rest, the rider starts pedaling and the bike begins to move forward. Think about what is actually pushing the bike and rider forward. To better think about this, consider what would happen if the bike started on ice. It would not be so easy for the biker to move forward. The biker turns the pedals, which in turn gets the wheels turning. If the bike is not on ice and the wheels turn on the ground without slipping, the bike moves forward. The wheels push *back* on the ground and

---

[4] I could, of course, use just *one* right triangle by taking only the first and last data points, but that model makes the terrain *too* simple and does not allow me to account for racing variations that occur on climbs and descents. Besides, computer speed is so fast that there is hardly a noticeable difference in time needed to compute the motion on one triangle versus many triangles.

the ground pushes *forward* on the wheels. This is explained by the same Newton's Third Law that we met in chapter 2 when we investigated how a football player got to, and then kicked, a football. For those of you new to physics, I hope by now you are seeing the true power of physical laws. A century ago, Einstein made the first of his two postulates of *Special Relativity* the idea that physical laws are the same in all "inertial reference frames" (for many phenomena, the surface of the Earth is, to a good approximation, an "inertial reference frame"). Understanding what makes footballs leap from our feet when we kick them and why bicycles move forward when we push the pedals allows us to build bridges and send satellites into space. Returning to the bike, recall that for every force there is an equal and opposite reaction force. The point is that the *ground* is responsible for exerting a force on the bike in the forward direction. As Newton's Third Law tells us, the force from the ground will not exist all by itself. The biker still has to turn the pedals to get the wheels to turn against the ground. There is no free lunch here.

There are other forces on the rider-bike combination. If the biker stopped pedaling on a steep uphill, you know exactly what would happen: the biker would come to a stop and then start to roll back down the hill. Conversely, if the rider descends a steep downhill, pedaling could stop and the bike would still fly down the hill. Clearly, a force is acting on the bike and rider. This force is gravity, which always pulls the rider-bike combination straight down toward the center of the Earth.[5] No matter whether the rider is moving uphill, downhill, on a straightaway, or even if the rider is sitting in a nice French tavern after the race, the Earth is always using gravity to pull the rider (and bike) straight down. As we saw in chapter 3, an object with mass m has weight mg. If we assume a mass of 77 kg for the rider-bike combination (a reasonable estimate, to judge from the cyclists' masses on the Tour de France web site), the weight force is about 755 newtons (or about 170 pounds).

We now have a forward force from the ground and a gravitational force. Let's look for more forces. Suppose the biker is sitting on the bike on level ground, so that there is no forward ground force as a result of pedaling. Let's ponder what it is that keeps the biker from falling to

---

[5] Keep in mind that the rider-bike combination is pulling *up* on the Earth with the same amount of force. That is what Newton's Third Law tells us.

the center of the Earth; there is, after all, still a gravitational force pulling the rider-bike combination straight down. We know the bike is sitting on the ground, so the ground must be exerting some other force on the bike. To think about what this force is, imagine pushing on a wall. If you are not too beefed up, you probably cannot push through the wall. If you push on the wall, the wall has to be pushing on you in the opposite direction. If the rider-bike combo pushes on the ground, the ground must push back up on the bike. This force is called the normal force, where the word "normal" means perpendicular to the ground.

It is easy to see that the force is perpendicular to the ground if the bike is sitting on level ground. To better understand why the force must be perpendicular to the ground if the ground is not level, imagine what would happen if the biker is sitting on one of our inclined planes and then all of a sudden the plane gets pushed up so that it becomes steeper and steeper. The normal force gets smaller and smaller until the plane becomes completely vertical, in which case the biker and bike fall to their doom. In that pernicious case, the plane can no longer help. Just keep in mind that as the incline gets steeper, the normal force decreases. Let's call the normal force F_N.

Now imagine that you are on a bike and flying down a hill. It feels great to have the wind hitting your face. The wind you feel is yet another force we have to consider on the rider-bike combo. The wind in one's face exerts a force, one we feel on us in the direction *opposite* to the one in which we are biking. The wind, or air resistance, resists the bike's motion. Even though the wind can feel great while we bike, the force from air resistance slows us down. This is, incidentally, a problem with cars (and a greater problem with trucks). When we move at a nice clip on the highway, a significant chunk of the gas we use goes into fighting air resistance. Of the roughly 35–40% of the gasoline that makes the pistons go up and down (this is the "thermal efficiency" of the engine), at least a quarter of it is used to fight air drag. Physicists have come up with a decent model for air resistance, one best summarized in the equation

$$F_D = \tfrac{1}{2} C_D \, \rho \, A v^2. \tag{4.2}$$

The force from the air is a *drag force*, hence the "D" subscript. The density of air is the lower-case Greek letter "rho," or ρ. Density plays a role because moving through water is tougher than moving through air; that is why $F_D \propto \rho$, where $\propto$ means "proportional to." Air density is around $1.2\,\text{kg/m}^3$ at sea level and room temperature.[6]

The A in equation (4.2) is the cross-sectional area of the rider-bike combo. To understand why $F_D \propto A$, think about when you are in a car moving along a highway and you stick your arm out the window. If your forearm and hand are in the "airplane" position, so that the area exposed to the wind is small, you feel a certain amount of resistance from the air. But when you turn your arm 90° so that the wind smacks directly into your palm, you notice greater air resistance. The point is that the more area there is, the more places there are for the air to whack into. So $F_D \propto A$.

The v^2 (where v is the bike's speed) in equation (4.2) is a little tougher to motivate. The idea is that if your arm is stuck out the window and your car's speed increases by a factor of two, the air resistance force you feel on your arm becomes *four* times larger. Try this in your car sometime (though I do not recommend doubling your speed once you hit 40 mph or so). The factor $\frac{1}{2}$ in equation (4.2) is there for a reason; the reason is, however, not so important for our discussion here.

Finally, the C_D factor is the drag coefficient and has no units whatsoever; it is a number. You can almost think of C_D as a type of "fudge factor," for if we know A, ρ, and v and determine $\rho A v^2/2$ and then notice that our result is a little different from some measured value of F_D, we simply clean it up with the C_D factor. In many situations, C_D depends on v. This may sound a little goofy, or even cavalier; it does turn out, however, that C_D does not vary *too* much for a rider-bike combo. Air resistance is an extremely complicated phenomenon to understand because the air hits the rider and bike in strange ways and sometimes swirls around behind them in complex patterns. Simply shaped objects like spheres and cylinders have different C_D values, even if they have the same cross-sectional areas. Scientists often determine the product $C_D A$ by measuring the drag force with, say, strain gauges, measuring ρ and v, and then solving equation (4.2) for $C_D A$. Typical values of $C_D A$ for a bicycle-rider combo are 0.25–$0.35\,\text{m}^2$, where the low end of that range might be about right for a cyclist crouching on the bike.

---

[6] Think about this for a moment, and be sure it makes sense to you. Air is not weightless. One cubic meter of air has a mass of about 1.2 kg. The weight of that air is thus about 12 newtons, or about 2.6 pounds.

Using equation (4.2) is thus a pretty good first step to including air resistance in our model. To give you a rough feel for the magnitude of the drag force, consider the 23.4 km/hr (or 6.5 m/s) average speed Armstrong had for stage 16 of the 2004 Tour de France. Plug in the numbers I've given you so far into equation (4.2) and we get

$$F_D \simeq \tfrac{1}{2}\,(0.30\,\text{m}^2)\,(1.2\,\text{kg/m}^3)\,(6.5\,\text{m/s})^2, \tag{4.3}$$

which yields $F_D \simeq 7.6$ newtons $\simeq 1.7$ pounds. That does not seem like much force, right? Remember now that stage 16 is straight uphill. How might the above number change for a cyclist zooming downhill? Speeds on steep downhills can be as great as three times the speed used in the calculation above. That means that the drag force is *nine* times larger than that found above, or about 68 newtons (15 pounds). Does *that* amount of force seem large? Imagine biking down a hill with the weight of a bowling ball pulling you backward. That might change your mind if you thought 68 newtons was not too much force. And imagine how much faster you would be flying if you were descending in a vacuum.

Just when you thought we had all the forces accounted for, there is one more to consider. Imagine that we roll a bike tire on a flat road. The tire can either stop eventually or keep rolling along in perpetuity. Even if there were no air around, the wheel would stop eventually, owing to friction. Where the bike wheel touches the ground, both the wheel and the ground experience some deformation,[7] and it is that deformation that keeps the wheel from rolling forever. This, by the way, is why you need to keep your tires inflated—not only on your bike, but on your car as well. The less deformation that takes place when the wheel is rolling, the less rolling friction that must be resisted. As with air resistance, a good chunk of our car's gasoline usage goes into overcoming rolling resistance. Of the gasoline that performs useful work to make the pistons go up and down, about an eighth of that is used to overcome rolling resistance. The harder that two surfaces push into each other, the greater the deformation, and, hence, the greater the rolling friction. We therefore take the rolling friction force, F_r, to be proportional to the normal force that we already discussed. In the form of an equation, we have

---

[7] The tire "smooshes" a little, whereas the road's deformation, if the road is made of a hard substance like concrete, takes place on a far smaller scale.

$$F_r = \mu_r F_N. \tag{4.4}$$

Another Greek letter, "mu" or μ_r (subscript "r" for rolling) is used as the dimensionless constant of proportionality. For a bicycle on a road, a common size of μ_r is about 0.003 (again, no units).

I have now told you about all the forces that we will consider on our rider-bike combo. I still need to give you an equation for the first force I described, the force due to the biker pedaling and having the tires push back against the road. The equation I want to use for the forward force, F_b, due to the biker pedaling is

$$F_b = P_b/v. \tag{4.5}$$

Here, P_b is the biker's power input and is measured typically in watts. Some of you will stare at equation (4.5) and wonder what happens if the speed is *low*. It may appear that I am claiming that the force from the ground gets HUGE. Equation (4.5) is effective only down to some low speed; call it $\tilde{v}$. To keep F_b from "blowing up" at small v (i.e. F_b gets large when v gets small), use $F_b = P_b/\tilde{v}$ for $v < \tilde{v}$, and use equation (4.5) for $v \geq \tilde{v}$.[8] What all this means is that if our biker is going slowly, he/she can put only a constant force on the pedals, or the steering will become erratic. That large force will eventually decrease once the bike gets moving faster than $\tilde{v}$. Think about trying to bike up a steep hill. You push really hard on the pedals just to move at a low speed. Once you start flying down the hill, you do not push nearly as hard on the pedals. For Tour de France modeling, I have found that $\tilde{v} = 6\,\text{m/s} \simeq 13.4\,\text{mph}$ works pretty well.

We now have to think about what to use for the biker's power input, P_b. Power is the rate at which energy is used with respect to time. A simplified way of thinking about power is

$$\text{power} = \frac{\text{energy}}{\text{time}}. \tag{4.6}$$

We know that our biker uses energy while pedaling through stage 16, and we know that cyclists have to *eat*. Estimating how much power is required to complete stage 16 is our task at hand. Say, for example, that we want

---

[8] This idea is discussed further in *Computational Physics* (2nd ed.) by Nicholas J. Giordano and Hisao Nakanishi (Pearson Prentice Hall, 2006), pages 24–25. This book also contains some simple ideas behind modeling the motion of a bicycle on a computer.

our cyclist to complete the race in 1 hour. We need only determine how much energy is used. There are many different forms of energy: mechanical, thermal, electrical, magnetic, chemical, nuclear, etc. In stage 16, our biker must start at 720 m above sea level and end up 1850 m above sea level. Imagine raising a book or stone or whatever you wish to raise. You actually give the object energy by raising it farther from the Earth's center.[9] Drop the object from the level of your waist and notice how it hits the ground. Then, drop it from the height of your head. It hits the ground a little harder when dropped from a greater height. The higher you raise the object, the more energy you give it. If we stay relatively close to the surface of the Earth, we can use an expression for the gravitational energy we have given our object, which involves the object's mass, the acceleration due to gravity, and something I will call Δh. Recall from chapter 2 that Δ means "change in," meaning Δh is the "change in height." If we end up at 1850 m after starting at 720 m, then $\Delta h = 1850$ m $- 720$ m $= 1130$ m. Our expression for the energy gained is $mg\Delta h$. When we multiply those three things together, we get an energy expressed in units of joules,[10] as long as m is in kilograms, g is in m/s^2, and Δh is in meters. With a mass, then, of our rider-bike combo of 77 kg,

$$\text{energy} = mg\Delta h = (77\,\text{kg})\,(9.8\,\text{m/s}^2)(1130\,\text{m}) = 852,698\ \text{joules.}\quad (4.7)$$

I have too many digits in my result above; I'll truncate them in a moment. We now use equation (4.6) to obtain an estimate of the power required to climb stage 16 in an hour (3600 seconds):

[9] Technically, you have changed the *potential energy* of the Earth-object system. The potential energy concept applies to more than one object. In other words, the notion of potential energy associated solely with just one object makes no sense. Given the fact that the Earth hardly moves when we raise an object in the air, we need not worry about this detail.

[10] The British unit of energy is the "foot-pound." One joule is about 0.74 foot-pound. Another common energy unit is the "calorie" (cal). One calorie is about 4.2 joules. The "kilocalorie" (kcal) is the "calorie" you see on food packages. Some call this unit the "nutritional calorie" or the "large calorie." As the prefix suggests, 1 kcal = 1000 cal. Your electric company charges you for the number of kW · hr (kilowatt-hours) that you use. A kW · hr is 3.6×10^6 joules. Finally, you may have heard of the BTU, or "British Thermal Unit," especially with respect to your home heating system. The BTU is the amount of heat (or energy) needed to raise one pound of water by 1°F. For that unit, 1 BTU $\simeq 0.25$ kcal $\simeq 1055$ joules. Note that I am not much worried about significant digits in this case!

$$power = \frac{852{,}698 \text{ joules}}{3600 \text{ seconds}} \simeq 237 \text{ watts}. \qquad (4.8)$$

Lance Armstrong actually completed stage 16 in 39′ 41″, which means that if we replace 3600 seconds in equation (4.8) with 2381 seconds, we get about 358 watts,[11] assuming of course that we are using the correct mass for Armstrong and his bike in equation (4.7).

After doing a quick calculation like the one above, a physicist needs to decide if the result is reasonable. Roughly three-and-a-half 100-watt light bulbs is our claim. Ah, but you have to wonder if we have not put the cart before the horse. If we wish to be good physicists and *predict* the time, then we certainly cannot use Armstrong's winning time to determine the power needed. We want a prediction in hand before the stage is even run. What we need to do at this point is consult the vast array of scientific literature out there and see what researchers have found for power output from elite cyclists.[12] Elite cyclists are capable of power outputs greater than Armstrong's 358 watts, at least for time periods of exercise under an hour. Lance Armstrong, too, is certainly capable of putting out more than 358 watts during stage 16.

If we increase our power estimate to an amount greater than 358 watts, you may wonder if Armstrong could finish stage 16 *quicker* than 39′ 41″. That would be true if gravity were the only force Armstrong had to fight on his ascent. As he bikes up the mountain, Armstrong battles air resistance, just like your car, and he battles rolling friction, again, just like your car. Not only must he output energy to overcome gravity, he must also output additional energy to overcome the frictional forces. Armstrong could easily put 475 watts or more into his biking to complete stage 16.

To keep the cart behind the horse, we'll take a quality power estimate from the available research. We then need to search the literature

---

[11] The British unit of power is the "foot-pound per second." The conversion is 1 watt $\simeq 0.74$ foot-pound/second. Another popular unit is the horsepower (hp). For that unit, the conversion is 1 hp $= 550$ foot-pound/second $\simeq 746$ watt.

[12] See, for example, *High-Tech Cycling: The Science of Riding Faster*, edited by Edmund R. Burke (Human Kinetics, Champaign, Illinois, 2003). This book contains many fine articles on the science of cycling. It covers many more aspects of cycling than I cover in this chapter.

some more for values of the other parameters. Or, if you have some suitable space along which to experiment, you could measure some of them yourself.

Model Results

Once we determine the necessary parameters, we are ready to use some mathematics and computational techniques. Since this is not a full-blown physics text, I'll spare you much of the detail. Newton's second law states that the vector sum of all the forces is equal to an object's mass times its acceleration. I have already described all the forces we need in our model; we now add them all up as vectors and equate that result to mass times acceleration. Getting some help from a computer, we solve the subsequent equations and determine the time it takes to go from start to finish in any of our Tour de France stages.

When I do the calculation for stage 16 of the 2004 Tour de France, I get 37′ 09″, which is 2′ 32″ too fast. But I just missed the actual total, by a little more than 6%. Given all that can happen in a Tour de France stage, and the complexity of trying to model an object as complicated as an athlete on a bicycle trying to climb a winding uneven hill, I think the prediction is not too bad. My calculated prediction is certainly better than the back-of-the-envelope estimates I made at the beginning. When a former student of mine, Benjamin Hannas, and I modeled the entire 2004 Tour de France, we predicted times for 20 of the 21 stages that were within 10% of the actual times.

For completeness, I want to back up that claim. I'm going to give you the parameters we chose in our model of the entire 2004 Tour de France and the results of our modeling. Below is a summary of our parameter choice.

$$m = 77\,\text{kg} \quad \text{(rider-bicycle mass)} \tag{4.9}$$

$$\rho = 1.2\,\text{kg/m}^3 \quad \text{(air density)} \tag{4.10}$$

$$\mu_r = 0.003 \quad \text{(coefficient of rolling friction)} \tag{4.11}$$

$$C_D A = \begin{cases} 0.25\,\text{m}^2, & \theta \leq 0° \quad \text{(downhill)} \\ 0.35\,\text{m}^2, & \theta > 0° \quad \text{(uphill)}. \end{cases} \tag{4.12}$$

That last entry in the above list, $C_D A$, combines the drag coefficient and cross-sectional area that I introduced you to in equation (4.2) into a single parameter. As I've discussed, people who measure drag forces on bicyclists will often quote values of the product $C_D A$. The reason we chose to use two different values for $C_D A$ is that a person biking downhill will tend to crouch on the bike and reduce the amount of area exposed to the wind. As you know from your own experiences on a bike, your body is fairly upright while you struggle to pump the pedals going uphill. The angle I gave, θ, is the angle of a given incline (see Figure 4.4).

The last parameter choice is the biker's power input. A top-notch athlete like a Tour de France cyclist is capable of putting out some hefty power and, furthermore, is capable of sustaining that power over much of a given leg of the race. As I mentioned earlier, there has been a great deal of research into the amount of power that world-class athletes can expend. That research, coupled with the multi-stage terrain of the Tour de France course, allowed us to estimate cyclist power outputs. We also had to consider that a cyclist is more likely to put out more power while struggling to get up a hill than during those nice wind-in-your-face downhill stretches. We determined all 498 angles that could be found for the entire 2004 Tour de France on the official web site (www.letour.fr) and then made an appropriate choice of power output for our model cyclist. Here is what we used:

$$P_b = \begin{cases} 200\,\text{W}, & \theta \leq -3.15° \\ 325\,\text{W}, & -3.15° \leq \theta < 3.55° \\ 425\,\text{W}, & 3.55° \leq \theta < 5.16° \\ 500\,\text{W}, & 5.16° \leq \theta. \end{cases} \tag{4.13}$$

Those angle cutoffs (given in numbers of degrees) look pretty strange, but when we looked at all those 498 angles, we found the above equations to be reasonable.

There is one more thing I need to tell you about modeling the Tour de France. If you look closely at the 21 stages of the race, four of them are short. For the 2004 race, stages 0, 4, 16, and 19 were the "time trials." Stage 0 is the "prolog" and takes less than 10 minutes to finish. We already talked about stage 16; stages 4 and 19 were each won in a little more than an hour. For these time trials, bikers will often wear more aerodynamic helmets and clothing. Even the bikes themselves are modified to reduce drag. Members of teams will often help each other by making use of "drafting," which occurs when one biker rides closely behind another so as to reduce wind resistance.[13] This practice can also be used while driving; many race car drivers make use of it during races. I would not, however, recommend doing this yourself, as it would mean you would have to tailgate the car in front of you to derive any benefit at all. Because cyclists make themselves a little more aerodynamic during time trials,[14] we reduced $C_D A$ by 20% for stages 0, 4, and 19. We also figured that because those stages were so short, the cyclists would put out more power, since they did not have to do so for more than an hour or so. We came up with the following values for the power output for stages 0, 4, 16, and 19:

$$P_b = \begin{cases} 200\,\text{W}, & \theta \leq -3.15° \\ 475\,\text{W}, & -3.15° \leq \theta < 5.16° \\ 500\,\text{W}, & 5.16° \leq \theta. \end{cases} \tag{4.14}$$

A Successful Model

I have now told you everything you need to do your own modelling of the Tour de France. It may seem as if I have given you a daunting array of information, equations, and parameter choices. You may scream that things are not that simple. Well, you're right. One does need to do a little research into what scientists have found out about athletes and bicycles. One also needs to look at the Tour de France terrain and use some physics. Think for a moment about what we have *left out*. The wind can also blow sideways, and we have not taken that into account. In the first third of the 2005 Tour de France, competitors experienced strong tail winds (around 16 km/hr,

---

[13] Although drafting is an important strategy in team time trials, it is forbidden in the individual time trials. A rider caught drafting in an individual time trial faces a time penalty.

[14] The exception here is stage 16. As we've already discussed, that stage was straight up a mountain, and nifty clothing and helmets would not help much there.

or 10 mph, with gusts upwards of 29 km/hr, or 18 mph), which significantly reduced stage-winning times. Bikers also bump into each other and sometimes crash, and we have certainly not introduced any "crash" equations. Those 498 angles may sound like a lot to examine for a race, but they fall short of describing the actual terrain. You may not know that Tour de France racers will eat and even use the "rest room" while racing along on their bikes. Suffice to say, we have not included *those* actions in our model. The point I'm trying to make here is that even though you may think that what I have described here is pretty complicated, it could be a *lot* worse.

Ben Hannas and I threw everything that I have described in this chapter into a computer and out popped Table 4.1. It shows the winning times for each stage, *not* the times for any individual biker like Lance Armstrong. His times do appear in the table if he happened to win the stage (like stage 16). We were trying to model the best of the best for each stage, which meant that we were after the winning times. We achieved this for every stage, except one, to better than 10%. (Forgive us for missing stage 17 by predicting a time nearly 12% too fast.) We failed to anticipate that it would be a foregone conclusion by the end of stage 16 that Armstrong would be on his way to victory, nor did we account for the fatigue that many racers must have felt with little rest between the end of stage 16 and the beginning of stage 17.

Some readers may say that I could have picked any set of parameters I chose until I got good agreement with the actual race. I could tweak five or six parameters until I hit the nail on the head. You would be (partly) right. There is, however, great evidence out there for the parameters we chose.[15] Hitting 20 out of 21 stages under 10% is more difficult than you might think. If we had, for example, upped the power so that one stage on which we were too slow would thereby improve, we would suffer worse

---

[15] Ben Hannas and I published two papers on Tour de France modeling. The first is "Model of the 2003 Tour de France," Benjamin Lee Hannas and John Eric Goff, *American Journal of Physics* 72 (5), 575–79 (2004). That paper introduced the incline-plane model for the Tour de France. Our second paper is "Inclined-plane model of the 2004 Tour de France," Benjamin Lee Hannas and John Eric Goff, *European Journal of Physics* 26 (2), 251–59 (2005). In addition to more details about Tour de France modeling in those two papers, you will find copious references for our parameter choices.

TABLE 4.1. Numerical results for the 2004 Tour de France

Stage	Actual winning time	Predicted time	Difference	% Difference
0	0 h 06′ 50″	0 h 06′ 51″	00′ 01″	0.24
1	4 h 40′ 29″	4 h 47′ 26″	06′ 57″	2.48
2	4 h 18′ 39″	4 h 38′ 33″	19′ 54″	7.69
3	4 h 36′ 45″	4 h 50′ 57″	14′ 12″	5.13
4	1 h 12′ 03″	1 h 14′ 49″	02′ 46″	3.84
5	5 h 05′ 58″	4 h 46′ 22″	−19′ 36″	−6.41
6	4 h 33′ 41″	4 h 29′ 23″	−04′ 18″	−1.57
7	4 h 31′ 34″	4 h 49′ 25″	17′ 51″	6.57
8	3 h 54′ 22″	3 h 56′ 28″	02′ 06″	0.90
9	3 h 32′ 55″	3 h 49′ 54″	16′ 59″	7.98
10	6 h 00′ 24″	6 h 00′ 01″	−00′ 23″	−0.11
11	3 h 54′ 58″	3 h 54′ 59″	00′ 01″	0.01
12	5 h 03′ 58″	5 h 29′ 24″	25′ 26″	8.37
13	6 h 04′ 38″	5 h 34′ 18″	−30′ 20″	−8.32
14	4 h 18′ 32″	4 h 29′ 54″	11′ 22″	4.40
15	4 h 40′ 30″	4 h 47′ 17″	06′ 47″	2.42
16	0 h 39′ 41″	0 h 37′ 09″	−02′ 32″	−6.38
17	6 h 11′ 52″	5 h 28′ 27″	−43′ 25″	−11.68
18	4 h 04′ 03″	4 h 04′ 25″	00′ 22″	0.15
19	1 h 06′ 49″	1 h 04′ 27″	−02′ 22″	−3.54
20	4 h 08′ 26″	3 h 44′ 09″	−24′ 17″	−9.77
Total	82 h 47′ 07″	82 h 44′ 38″	−02′ 29″	−0.05

Note: The difference column 4 gives the difference between column 3 and column 2, whether plus or minus. The % difference column 5 is [(column 3 − column 2)/(column 2)]× 100%. The actual winning times were taken from www.letour.fr.

results on stages where we were too fast. We were more interested in using what is in the literature than in playing parameter-tweak games.

There may still be a skeptic or two out there who knows that, in science, one needs to apply a model to more than one "experiment." That is an excellent objection. So Ben and I took the model we created for the 2004 Tour de France and ran it for the 2003 race. Aside from noting that the short time trials in 2003 were done in stages 0, 4, 12, and 19, everything else is the same. I put the necessary angle data from www.letour.fr into the computer and let 'er rip. Table 4.2 shows the results. We missed three stages by more than 10%, and although we missed the sum of the all the stage-winning times by almost an hour and a half, we were off by just under 2%. I'm inclined to think that's pretty good. When applied to two Tour de France races, our 2004 model predicted 38 of the 42 combined stages to an accuracy of better than 10%.

TABLE 4.2. Numerical results for the 2003 Tour de France using 2004 model

Stage	Actual winning time	Predicted time	Difference	% Difference
0	0 h 07′ 26″	0 h 07′ 42″	0 h 00′ 16″	3.59
1	3 h 44′ 33″	3 h 55′ 39″	0 h 11′ 06″	4.94
2	5 h 06′ 33″	4 h 50′ 28″	−0 h 16′ 05″	−5.25
3	3 h 27′ 39″	3 h 58′ 57″	0 h 31′ 18″	15.07
4	1 h 18′ 27″	1 h 16′ 43″	−0 h 01′ 44″	−2.21
5	4 h 09′ 47″	4 h 44′ 07″	0 h 34′ 20″	13.75
6	5 h 08′ 35″	5 h 20′ 51″	0 h 12′ 16″	3.98
7	6 h 06′ 03″	6 h 00′ 58″	−0 h 05′ 05″	−1.39
8	5 h 57′ 30″	5 h 49′ 13″	−0 h 08′ 17″	−2.32
9	5 h 02′ 00″	4 h 47′ 44″	−0 h 14′ 16″	−4.72
10	5 h 09′ 33″	4 h 55′ 39″	−0 h 13′ 54″	−4.49
11	3 h 29′ 33″	3 h 45′ 07″	0 h 15′ 34″	7.43
12	0 h 58′ 32″	0 h 56′ 34″	−0 h 01′ 58″	−3.36
13	5 h 16′ 08″	5 h 15′ 37″	−0 h 00′ 31″	−0.16
14	5 h 31′ 52″	5 h 24′ 51″	−0 h 07′ 01″	−2.11
15	4 h 29′ 26″	4 h 27′ 43″	−0 h 01′ 43″	−0.64
16	4 h 59′ 41″	5 h 01′ 35″	0 h 01′ 54″	0.63
17	3 h 54′ 23″	4 h 12′ 08″	0 h 17′ 45″	7.57
18	4 h 03′ 18″	4 h 49′ 56″	0 h 46′ 38″	19.17
19	0 h 54′ 05″	0 h 55′ 04″	0 h 00′ 59″	1.82
20	3 h 38′ 49″	3 h 24′ 58″	−0 h 13′ 51″	−6.33
Total	82 h 33′ 53″	84 h 01′ 34″	1 h 27′ 41″	1.77

Note: The difference column 4 gives the difference between column 3 and column 2, whether plus or minus. The % difference column 5 is [(column 3 − column 2)/(column 2)]× 100%. The actual winning times are taken from www.letour.fr.

We have applied our model to other tour races, as well, including the 2005 Tour de France. To apply the model described here to the 2005 race, we had to model the effect of the strong tailwinds. We did this by increasing the power input, thus making an "effective" power input that includes effects from the biker and the tailwind. We had just two stage predictions come in over 10%.

What I hope that I have been able to convince you of in this chapter is the virtue taking a problem that is complicated and simplifying it as much as possible. If the reduction proves to be too simple, we introduce a little more complication. The balance between making a model simple and making it accurate is the struggle that we physicists always face.

Let's now jump off the French Alps and move on to one of the greatest leaps in human history.

5

A Leap into the Unknown

More Projectiles and Angular Momentum

A Beamonesque Moment

On May 29, 1953, Sir Edmund Hillary and Tenzing Norgay ascended to the top of Mount Everest and stood on top of the world, where no human had ever stood.[1] Millions watched on July 20, 1969, as Neil Armstrong hopped off a ladder onto the surface of our Moon. No human had ever stepped on the Moon before that indelible moment. Sprinkled throughout human history are moments like these.

Science has also had its share of "firsts," where individuals reached a plane of achievement that no other human had visited. You have probably read or heard about several at some point or other in your life. I have always enjoyed the story of how David Lee, Douglas Osheroff, and Robert Richardson discovered superfluid Helium 3 in the early 1970s. They won the 1996 Nobel Prize in Physics for their work, which spectacularly demonstrated quantum effects. The story of Andrew Wiles proving Fermat's last theorem as the twentieth century came to a close is another classic.[2]

Many "firsts" have been achieved because people were trying to be the "first" to do something. Hillary and Norgay knew they would put the first human footsteps on the summit of Everest. Neil Armstrong knew

---

[1] There is a debate whether or not George Mallory and Andrew Irvine reached the summit of Mount Everest in 1924 before perishing. Did they die on the way up or on the way down?

[2] I highly recommend watching the *NOVA* "Adventures in Science" piece entitled *The Proof* (1997).

he would be the first to step on the Moon. On other occasions, the individuals in pursuit have failed, perhaps because they may not have known exactly what path to follow, or even whether they would recognize a breach in the vaunted plane of novel achievement. Lee, Osheroff, and Richards initially thought their discovery was of a new phase of *solid* Helium 3. In as quintessential an example as there is for the importance of university tenure, Wiles spent eight years of his life in near seclusion trying to best Fermat. He did not know he would be successful; in three hundred years no human had succeeded doing what Wiles eventually did.

What Bob Beamon did in the long jump at the 1968 Summer Olympics in Mexico City was to reach a plane of human achievement so far beyond all that had previously been reached that his competition simply surrendered on the spot. Lynn Davies, who had taken home the gold in the long jump for Wales at the 1964 Summer Olympics in Tokyo, remarked to Beamon, "You have destroyed this event." Beamon's jump was so amazing that a new adjective, *Beamonesque*, popped into our vernacular.

Before Beamon's famous jump, the world record long jump, 27 feet, $4\frac{3}{4}$ inches (8.35 m), was held jointly by the American Ralph Boston and the Russian Igor Ter-Ovanesyan. Beamon, whose previous best jump was 27 feet, 4 inches (8.33 m), jumped 29 feet, $2\frac{1}{2}$ inches (8.90 m). Beamon nearly jumped what would have been a first down in football (10 yards)! Had Beamon leaped from the top of the three-point arc on a National Basketball Association court, he would have landed *behind* the basket, nearly a foot and a half *out of bounds*. Think about that for a moment. Given the enormous amount of training that goes into Olympic long jumping, and given the specific task of trying to jump as far as one can during a long jump event, I doubt that any human, anywhere, had ever jumped farther than 27 feet, $4\frac{3}{4}$ inches before Beamon made his jump. I suppose it is conceivable that someone had jumped a tad farther in practice, and who really knows how well cavemen jumped (probably not far, given their body structure). What Beamon did was extend the world record by not just a "tad farther," as is usual in record-breaking performances, but by $21\frac{3}{4}$ inches farther. That is 1 foot, $9\frac{3}{4}$ inches (0.552 m), or about 6.6%, farther than any human had ever jumped, and almost 6.9% farther than Beamon himself had ever jumped. While still in the air, Beamon soared through both the 28-foot *and* 29-foot makers. A steel tape measure was used to determine Beamon's jump distance because the optical device

Figure 5.1. An optical measuring device used to determine the length of a long jump. Note the rod in the sand, which marks the jumper's landing point. The small black ball at upper left is the top of the rod.

(see Figure 5.1) employed at the time for determining jump measurements fell off its rail before it reached Beamon's landing point. Ter-Ovanesyan said, "Compared to this jump, we are as children."

Physics Time

To model Beamon's epic jump, let's do what we did with Doug Flutie's Hail Mary pass. Equation (3.10) is an equation for the range of a projectile. Let's try that. Recall that equation (3.10) is

$$R = \frac{v_0^2 \sin 2\theta}{g}, \qquad (5.1)$$

where v_0 is the launch speed, θ is the launch angle, and g is the magnitude of the acceleration of gravity. For the launch speed, assume that a world-class athlete like Bob Beamon could run 10 m/s at his launch point.

Even if Beamon was not able to run 100 meters in 10 seconds, his *instantaneous* speed at the launch point could have been about 10 m/s. I will also assume that Beamon jumped at the angle that maximizes range, namely $\theta = 45°$. But if I plug everything into equation (5.1), with $g = 9.8\,\text{m/s}^2$, I get $R \simeq 10.2$ m (or about 33.5 feet). We are in the ballpark, but we're about 15% too long. We should be able to do better than that.

If I plug Beamon's actual range of 8.90 meters into equation (5.1) and solve for the launch angle, I get $\theta \simeq 30.4°$. That is a far cry from 45°. So, did Beamon jump at 45° to maximize R in equation (5.1)? Did he jump at a smaller angle, like the 30.4° just obtained, and get less range? Or, is there a problem with our model?

To answer those questions, we need more information about Beamon's jump. I watched the jump a few dozen times and found that his time of flight was nearly 1.0 s. That is as accurately as I could measure Beamon's flight time with my stopwatch. As with Flutie's Hail Mary pass, we know Beamon's range, R, and time of flight, T. We will thus use the following:[3]

$$\tan \theta = \frac{gT^2}{2R} \tag{5.2}$$

and

$$v_0 = \sqrt{\left(\frac{gT}{2}\right)^2 + \left(\frac{R}{T}\right)^2}. \tag{5.3}$$

Inserting $R = 8.90$ m and $T = 1.0$ s into the above equations gives $\theta \simeq 28.8°$ and $v_0 \simeq 10.2$ m/s. The angle is significantly different from our initial 45° guess; the launch speed is, however, close to our 10 m/s estimate.

We have now plugged several numbers into a bunch of equations and we've come up with a few different angles, launch speeds, and ranges. You have to wonder if anything we have calculated is correct. We know that Beamon's jump was 8.90 m, and we can estimate with a stopwatch that his flight time was about 1.0 s. The rest of the numbers we've calculated are suspect, and for good reasons. I mentioned in footnote 17 in chapter 3 that using equations in books requires us to think carefully about the *assumptions* that go into creating those equations. We simply

[3] See footnote 18 in chapter 3.

cannot go willy-nilly into a book and pluck out an equation that happens to employ parameters we know (like R and T) and one we want (like v_0). In addition to assuming that Beamon's acceleration was constant while he was off the ground (as seems reasonable enough), we ignored air resistance. Even worse, all of the equations we have used make the assumption that Beamon's initial and final heights were the same. A closer look at the video reveals that compared to where he was when he leaped, Beamon was actually *lower* to the ground when he landed. A more subtle assumption I made in chapter 3 is that the equations used there apply to the motion of a point particle. That was a reasonable assumption for a football that did not change its shape while in motion. Beamon's body, however, like anyone's engaged in a long jump, was doing all kinds of crazy stuff while in the air. We need to improve our model by taking all of the aforementioned effects into account.

What's the Point?

We all agree that Bob Beamon is not a point particle; Mr. Beamon could surely testify to that fact. Does that mean that everything we did in chapter 3 is now useless? Luckily, the answer to that question is "no." It turns out there is something within Bob Beamon that effectively moved like a point particle during his jump. That something is the location of his center of mass. When we study an object's *translational* motion, the center of mass is the point where we may imagine all of the object's mass to be concentrated. In other words, we imagine that there is the center of mass of an object experiencing changes in its translational motion due to the presence of external forces. Newton's second law, given by equation (3.6) as $m\vec{a} = \vec{F}^{\text{net}}$, allows us to determine the trajectory of an object's center of mass. All of the crazy stuff Beamon did in the air with his arms and legs is something we will consider soon in the context of *rotational* motion, but none of Beamon's arm and leg motions affected the motion of his center of mass. At the instant Beamon left the Earth's surface, his center of mass can be described as if it were a single point particle moving under the influence of a gravitational force, air resistance, and the wind. Technically speaking, the gravitational force acts effectively at what is called the *center of gravity*. Since the gravitational force from the Earth on a mass particle within an object varies like the inverse-square of the distance between the particle and the Earth's center of mass, a proton in Beamon's foot actually feels an ever so slightly larger gravitational force than a proton in his head. Beamon's center of gravity is thus slightly lower than his center of mass. It turns out that Beamon's center of gravity is about $\frac{1}{400}$ the width of a human hair beneath his center of mass. So let's forget about the distinction between center of gravity and center of mass!

The center of mass is not a real mass particle in an object. It is an imagined point, which we might think of as the average position of all the points of an object's mass. The center of mass of an object need not be *inside* an object. Figure 5.2 shows a confectionery delight. Where do you think the center of mass of the doughnut is located? Kudos if you said the doughnut's center of mass is in the middle of the hole.

Non-sports athletes like ballet performers use center of mass ideas to create some exciting moves on stage. During a grand jeté, a ballerina leaps up off the stage and immediately lifts her arms and legs upward. Her center of mass moves slightly upward, meaning that her head does not go as high as it would have gone had she not raised her arms and legs. Her

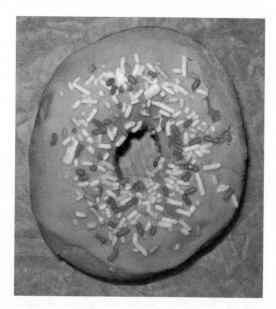

Figure 5.2. As Homer Simpson would say, "Mmmmmm ... doughnut!"

Figure 5.3. Lynchburg College athlete Ashley Palmer executes the Fosbury Flop. (Photo by John Shupe. Reprinted by permission from Lynchburg College.)

head thus does not move much while she is in the air; some might think she "hangs" in the air. Track and field athletes also make use of the center of mass idea. Figure 5.3 shows an athlete executing the "Fosbury Flop" while high jumping. The technique was named after Dick Fosbury, who had won the gold medal in the same 1968 Summer Olympic games in which Bob Beamon excelled. The idea is for the jumper to pass over the bar while his or her center of mass passes at or below the bar. Since the maximum height of the center of mass is constrained by conservation of energy once the jumper leaves the ground, turning the body around the bar allows for maximum bar height.

When Beamon jumped, he moved his arms and legs in such a way that his center of mass was *lower* at his landing than at his launch. In chapter 3, we focused on projectile motion for which the initial and final heights were the same. This is not how to approach Beamon's jump. Consider the trajectory of Beamon's center of mass if there were no air around. Equation (3.12) gives the vertical position as a function of horizontal position, assuming $y = 0$ when $x = 0$. Figure 5.4 shows a graph of what Beamon's center of mass trajectory might have looked liked as it progressed across his famous jump. To make the graph, I employed a range of 8.90 m and a time of flight of 1.0 s. I further took Beamon's

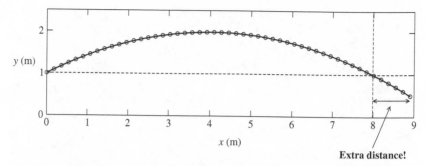

Figure 5.4. Trajectory of Bob Beamon's jump as if in vacuum. The open circles show Beamon's center of mass at 0.02-second intervals. The vertical dashed line illustrates where Beamon would have landed had he not altered the configuration of his body while in flight.

center of mass to be 1.0 m off the ground when he jumped and 0.5 m off the ground when he landed, which represents a simple shift in origin from the one assumed in equation (3.12).[4] I then solved[5] equations (3.7) and (3.8a) for v_0 and θ and found $v_0 \simeq 9.93$ m/s (about 32.6 ft/s or about 22.2 mph) and $\theta \simeq 26.3°$. The launch speed is consistent with our earlier estimates. The launch angle, however, is the smallest angle we have determined. This calculation demonstrates how poor our initial 45° guess was.

Look again at Figure 5.4. Beamon's center of mass is almost 2 m off the ground at the trajectory's maximum height. Note also that Beamon's center of mass at the landing point is indeed half a meter below where it was at the launch (horizontal dashed line). Had he done nothing with his body to allow him to land with his center of mass lower than where it started, Beamon would have landed at the location of the vertical dashed line, around 8.0 m (or about 26 ft 3 in). That would have been a full 0.9 m (nearly 3 ft) short of his record jump. He would also have been 0.35 m (almost 14 in) short of the world record. There is clearly a huge advantage to landing with one's center of mass well below its initial launch height. How did Beamon (and other long jumpers) do it?

---

[4] The center of mass locations are merely my estimates. I cannot determine the *exact* location of Beamon's center of mass. Though my estimates could be 10% off (or thereabouts), a drop of half a meter is sufficiently accurate to allow us to continue the calculation.

[5] Recall that $v_{0x} = v_0 \cos\theta$ and $v_{0y} = v_0 \sin\theta$.

The Beauty of Angular Momentum Conservation

A beautiful aspect of the universe's physical laws is that those laws constrain what objects can do. If a high jumper leaps into the air, energy conservation puts constraints on how high the jumper's center of mass can go. The jumper's kinetic energy loss cannot exceed his or her gravitational potential energy gain. You probably have a familiarity with energy conservation even if you never knew what to call it. If you throw a ball in the air, perhaps to practice catching pop flies, you know that you have to throw the ball harder if you wish it to go higher. In other words, you have to supply the ball with more energy if you want to throw it higher.

Conservation of linear momentum (more on this later!) helps us reconstruct accidents involving automobiles. It also gives us a clue to what is going on during a football tackle. Conservation of electric charge allows us to keep track of current in all of our electronic devices. Conservation of angular momentum will tell us how Bob Beamon got his center of mass to a lower height in the pit than it was at the launch point.

Because conservation laws constrain what can happen to an object, we are able to make predictions. After all, if energy was not conserved, who's to say that if you threw a ball up in the air that it would not keep going until it hit the Moon? Much of how conservation laws constrain what happens in nature is so hard-wired into us that if we were to see a thrown ball keep going up and up and up until it left our sight, we would surely think, "Gee, that's odd!" Even without a shred of physics knowledge, people have a sense of what should happen when everyday objects are set in motion.

But understanding how Beamon got his center of mass lower in the pit than at the launch may not be so hard-wired in us. Recall that once Beamon left the Earth's surface at the launch, his center of mass motion was determined by gravity, air resistance, and wind. The only way Beamon could change his center of mass motion was to alter his cross-sectional area and thereby change the drag force on him. (See equation (4.2) in the previous chapter.) Beamon had no control over the Earth's gravitational pull. Let's ignore the effects from the air for now and think about what Beamon could have done to produce a jump that accords with the trajectory shown in Figure 5.4. Even if he could do nothing in the air to alter the parabolic shape of his trajectory, Beamon *could* do something about how his body was oriented about his center of mass. To understand how he did it, we need to think about *rotational motion*.

An object's center of mass moves according to Newton's second law, given by equation (3.6). The rotational analog of Newton's second law is given by

$$I \vec{\alpha} = \vec{\tau}^{\text{net}}, \tag{5.4}$$

where I is an object's moment of inertia, $\vec{\alpha}$ (lower case Greek letter "alpha") is its angular acceleration (a vector), and $\vec{\tau}^{\text{net}}$ (lower case Greek letter "tau") is the net external torque (also a vector) acting on an object. The three quantities (I, $\vec{\alpha}$, and $\vec{\tau}^{\text{net}}$) in the above equation are analogous to the three quantities (m, $\vec{a}$, and $\vec{F}^{\text{net}}$) in Newton's second law for translations. An analogy, however, is as far as we can go. The symbols in the two equations are not the same; they do not even possess the same units.

An object's moment of inertia is essentially a description of how that object's mass is distributed about some rotation axis. Formally, I is defined as

$$I = \sum_i m_i r_i^2, \tag{5.5}$$

where m_i is the mass of the ith particle in an object and r_i is the perpendicular distance (i.e. shortest path) from that mass particle to the rotation axis.[6] The capital Greek "Sigma," Σ, is the mathematical notation for summation. Take the spinning Earth as an example. The rotation axis is a line that passes through the north and south poles; the Earth spins about that line. To compute I of the Earth about that rotation axis,[7] we would need to get the mass of each particle in the Earth and measure its perpendicular distance from the rotation axis. Then we would sum up all the mass times distance squared terms as specified by equation (5.5).

Any idea how many particles there are in the Earth? What do I even mean by "particles" anyway? Chunks of rock? Grains the size of dust? Are we talking atoms here? Or, are we talking about protons and neutrons?

---

[6] Formally, equation (5.5) is a special case of a more general result involving what is known as the inertia *tensor*. The general result is well beyond the scope of this book.

[7] Note that we are free to choose any other line to compute I. The one passing through the geographic poles is the most natural choice. Be aware, though, that I is defined only with respect to a chosen axis.

Or, heaven forbid, quarks and gluons? From the inner core of the Earth all the way up to the Earth's crust, the mass density and chemical composition vary. Estimating the number of atoms is challenging. The Earth's mass is about six trillion trillion kilograms ($\sim 6 \times 10^{24}$ kg). Protons and neutrons have a mass of about 1.67×10^{-27} kg, meaning there are close to 4×10^{51} protons and neutrons making up the Earth. That is an insanely big number. We have neither the desire nor the ability to deal with that many terms in equation (5.5). And nobody wants to pretend the Earth's particles are kilogram chunks, since we would then have to add up six trillion trillion separate terms.

Computing I is thus often a challenging endeavor. Fortunately, we can make assumptions and use calculus[8] to give us good estimates of I for various objects. Calculus lets us show, for example, that for a uniformly dense sphere spinning about a line passing through its center, $I = 2MR^2/5$, where M is the sphere's mass and R is its radius. Although the Earth is not a uniform sphere, we can use the aforementioned result to estimate the size of Earth's I. With the Earth's radius about 6.4×10^6 m, $I \sim 10^{38}$ kg·m^2. That is not the exact value of Earth's I; it does, however, give the right order of magnitude. Besides, it's a lot easier to calculate that number than the hyper-Beamonesque effort of adding up six trillion trillion terms.

Enough about the Earth; it was my example to give you a flavor for what I is and how to calculate it. The main qualitative point here is that if you move mass away from a rotation axis, you increase the size of the moment of inertia. Likewise, if you move mass closer to the rotation axis, you decrease the size of I. Stand up and then imagine a vertical line passing through your head and between your shoes. Now, raise your arms so that they are parallel to the ground (perpendicular to your torso). You have just increased the moment of inertia of your body about that vertical line.[9]

We still need to discuss two more terms from equation (5.4). The quantity analogous to linear acceleration is $\vec{\alpha}$, the angular acceleration. Like its linear counterpart, $\vec{\alpha}$ represents the time rate of change of a velocity. The velocity in question is the angular velocity. Physicists love to use Greek letters for several of the angular quantities, and we call the angular velocity $\vec{\omega}$, using the lower-case Greek letter "omega" (a vector). The average

---

[8] The sum in equation (5.5) must be converted into an integral.
[9] This idea will be incredibly important in the next chapter.

angular speed, i.e., the magnitude of the average angular velocity vector, is given by

$$\omega_{\text{ave}} = \frac{\Delta\theta}{\Delta t}, \tag{5.6}$$

where the quantity analogous to linear displacement is the angular displacement, $\Delta\theta$. The angle θ is measured in radians in SI units.[10] Compare equation (5.6) with equation (2.1) and note that I am not using vector notation here.[11] The average angular acceleration is

$$\alpha_{\text{ave}} = \frac{\Delta\omega}{\Delta t}, \tag{5.7}$$

where α is measured in rad/s^2.

Instantaneous quantities θ, ω, and α are well defined, too.[12] Directions are well defined for ω and α; however, the directions are less intuitive than those associated with linear quantities like $\vec{v}$ and $\vec{a}$. Imagine swinging a baseball bat or a golf club. In addition to displacing its center of mass, you need to rotate the bat or club. You need at least a two-dimensional plane for that rotation, and the rotation is complicated enough that there is no single plane of rotation. Even a more simplified example like a rotating DVD must have two dimensions in which to execute its rotation. It is difficult to imagine picking some direction in that two-dimensional plane of the rotating DVD as the one and only direction of rotation.

What we do instead is use a very primitive device known as your right hand. If your fingers curl with the rotation, your thumb points in the direction of $\vec{\omega}$. The direction of $\vec{\omega}$ is thus perpendicular to the plane of rotation. See Figure 5.5 for an example of how to find $\vec{\omega}$ for a rotating bicycle wheel. Note that rotations are a little more complicated than linear translations because of the added dimension needed to describe the motion.

---

[10] In the SI unit system, the radian is formally "derived" from one of the seven SI base units. In reality, all that means is that no dimensional units are associated with radians. If you multiply a radian by a meter, for example, you get a meter. As for conversions, there are 2π radians (abbreviated "rad") in a complete circle. Since there are 360 degrees in a complete circle, 1 rad $\simeq$ 57.3°. Some engineers use the "grad" as their angle unit. There are 400 grads in a complete circle, meaning 1 rad $\simeq$ 63.7 grads.

[11] For mathematical reasons beyond the scope of this book, we cannot write a finite angle as a vector. The "direction" part of a vector is not well defined for a finite angle. It turns out that the direction *is* well defined if the angle is infinitesimal, i.e. very small.

[12] We must use calculus for the instantaneous quantities, meaning $\omega = d\theta/dt$ and $\alpha = d\omega/dt = d^2\theta/dt^2$.

Figure 5.5. The bicycle wheel spins counterclockwise. The fingers of my right hand curl in the direction of the rotating wheel, while my thumb gives the direction of $\vec{\omega}$. (Photo by Jim Carico.)

The last term of equation (5.4) we need to discuss is the net external torque, $\vec{\tau}^{\text{net}}$. Torque is the rotational analog of force. Suppose you wish to open a door. There are many ways you could push or pull on the door. Where is the best place to apply the force from your hand? What is the best direction for that force? Have you ever wondered why most doorknobs are located near the *edge* of the door, as far from the hinge as possible? The answers to all these questions lie in an understanding of torque.

The basic definition of torque is the product of a force with a lever arm distance. A lever arm distance is defined as the perpendicular distance from a chosen rotation axis to the point where the force is applied. Like moment of inertia, torque is defined in terms of a chosen axis; we speak of the "torque about an axis of rotation." If you specify a different axis, you get a different torque (in general). In many cases, there is one rotation axis that begs to be chosen. The line passing through the center of the bicycle wheel and parallel to my thumb in Figure 5.5, for example, is the most natural choice of rotation axis.[13]

Note how important the lever arm definition is to generating a large torque. I get a lot of torque if I rotate the bicycle wheel in Figure 5.5 in such a way that my hand force is tangent to the wheel. In that case, the lever arm distance is the radius of the wheel. If, however, I apply my hand force at the same point on the wheel, but instead push along a line passing through the center of the wheel, I generate no torque. The perpendicular distance between the line of action of my applied force and the rotation axis is zero in that case. This is why you pull a door open by pulling in a direction perpendicular to the face of the door. For if you pulled such that your force's line of action passed through the hinge, you could not rotate the door.

Incidentally, have you ever seen those fancy doors with the doorknob placed right in the middle of the door? The gain in opulence is quickly lost when one tries to open the door. To produce the same amount of torque as someone pulling on the end of the door, one must exert twice the force in the middle of the door, since the lever arm distance is half what it is for the person pulling at the end. This is, by the way, why wrenches (or so-called torque wrenches) have long arms; the desire is for a long lever arm distance. Pulling at the end of a wrench requires about 5% of the force one needs to turn a nut by applying one's hands directly to the nut.

I have taken a couple of pages to go through the rotational quantities as carefully as I could because, as I mentioned before, they are not as intuitively familiar to us as their corresponding translational quantities are. Ok, we have but one more rotational quantity to consider before getting back to Beamon's jump. It turns out that the net external torque on an object is related to the time rate of change of a quantity called an object's *angular momentum*, which I'll call $\vec{L}$.[14] What that means is

---

[13] Please keep in mind that nature couldn't care less what nerdy physicists choose for a rotation axis. The "natural choice" is usually one that not only seems obvious, but is the one for which the mathematical description is often the simplest.

[14] A rigorous proof of $d\vec{L}/dt = \vec{\tau}^{\text{net}}$ is beyond the scope of this book.

that we have an opportunity to use a conservation law. For if the net external torque on an object is zero, that object's angular momentum cannot change in time. The true power in that statement is that if we know an object's angular momentum at one point in time, we know what the angular momentum has to be at any other time.

Do we really need to know how to work with the "time rate of change of angular momentum?" Fortunately, nature provides us with an alternate way of writing angular momentum, namely

$$\vec{L} = I\,\vec{\omega}. \tag{5.8}$$

Note that $\vec{L}$ is a vector that points in the *same* direction as $\vec{\omega}$. In Figure 5.5, $\vec{L}$ points in the direction of my thumb. If the net external torque is zero, $\vec{L}$ is unchanged in both its magnitude and its direction. That does *not*, however, mean that I and $\vec{\omega}$ must each be fixed. This simple idea explains a lot of what we see in the wide world of sports!

Before we turn back to Beamon, I'll end this rather long section with an example. Figure 5.6 shows me on a platform that can spin. I am holding a 1-kilogram mass in each hand. My wife got me spinning; i.e., she applied an external torque to me. In Figure 5.7, I have pulled my hands close to my body's center, i.e. close to the vertical axis of rotation. I hope you can tell by the blurry photo that I am spinning faster (and getting dizzy!). While the net external torque on me is not exactly zero, owing to friction with the platform and the axle, we can use the idea of angular momentum conservation over a reasonably short time interval (the time it takes me to pull my arms in). Even if the thought of calculating my moment of inertia scares you as much as it scares me, we can understand why I spun faster upon bringing my arms in. What happens to my moment of inertia when I bring those kilogram masses close to my chest? Equation (5.5) tells us that the mass particles I moved now have smaller distances to the rotation axis, meaning I decreases. If the product of moment of inertia and angular velocity must be fixed, my angular velocity had to increase. It's that simple!

Back to Beamon

We are now in a position to understand how Bob Beamon was able to land in the pit with his center of mass lower than its location at the launch

Figure 5.6. I am holding a 1 kg mass in each hand while spinning on a platform. Note my outstretched arms. Exposure time was 0.2 s. (Photo by Jim Carico.)

point. Look at Figure 5.8. In that glorious photo, we see Beamon in a much different orientation than the upright running posture he had been in at takeoff. Note that with his legs extended forward, Beamon can land in the pit with his center of mass significantly lower to the ground than if he were upright. Since Beamon's measured distance is determined by the point where the backs of his heels hit the pit (unless he lands on his butt or his hands), having his legs forward is an added plus.

Figure 5.7. After bringing my arms in, I spin faster. Exposure time was again 0.2 s. (Photo by Jim Carico.)

How did Beamon get his legs out in front of him, nearly parallel to the ground? Angular momentum conservation answers the question! While in the air, gravity is the dominant force on Beamon. If we imagine a rotation axis parallel to the ground, pointing right to left in Figure 5.8, and passing through Beamon's center of mass, then gravity cannot exert a torque about that axis, since the gravitational force has no lever arm with that axis. If we ignore the small effects of air resistance for the moment, there are no other forces on Beamon while he is in the air. That means that there

Figure 5.8. Beamon is halfway to immortality. Note how his arms are thrust mostly backward while his legs are thrust forward. (© Bettmann/CORBIS)

are no external torques acting on Beamon while in flight. Beamon's total angular momentum about the aforementioned rotation axis is therefore *zero*.

You might say, "Gee, if Beamon's total angular momentum is zero, how is that going to help us?!?" The beauty of conservation laws, and conservation of angular momentum in this case, is precisely that they constrain what can happen. It is certainly true that Beamon can do nothing in the air to exert a net torque on himself that causes him to rotate about our chosen rotation axis. Look at Figure 5.8 again and notice where Beamon's arms are located. While in air, he has thrust his arms as far behind him as he possibly can. During his qualifying jump the day before, Beamon's arms were thrust between his legs. While that does help get his legs forward, thrusting his arms out to his side allows his arms to get farther behind his body. Note that Beamon's elbows and fingers are locked, thus extending his arms fully, meaning, by virtue of equation (5.5), that his arms' moment of inertia is maximized. Beamon's torso, thanks to some amazingly strong abdominal muscles, is rotated forward. Think back to the right-hand rule we discussed earlier. Beamon's torso going forward and his arms going backward create a rotational velocity vector that points to the *right* in Figure 5.8. Remember to swing the fingers of your right hand in the direction of either his arms or his torso. Your right thumb then points to the right. But that means that Beamon's torso and arms have contributed angular momentum to the right (during the time they were in motion), by virtue of equation (5.8). Since Beamon's *total* angular momentum must be *zero* at all times while in flight, something else on Beamon's body must rotate in such way as to provide a compensating angular momentum to the *left*. You guessed it! Beamon's legs are thrust forward while his arms and torso are rotated. Using the right-hand rule, you can see in Figure 5.8 that moving Beamon's legs forward means a rotational velocity vector pointing to the left (only while the legs are in motion). Swing the fingers of your right hand forward with Beamon's legs; your right thumb now points to the left. Using equation (5.8) again, we see that Beamon's legs contribute angular momentum to the left (again, only while the legs are rotating about Beamon's center of mass).

Once Beamon's arms and torso have stopped rotating about his center of mass, his legs stop rotating, too. His total angular momentum is *still* zero. But look what happened! Beamon's body is now folded in such a way as to allow his center of mass to almost make it to the ground. Recall Figure 5.4 and the distance gained by lowering the center of mass at the landing point. Additionally, Beamon's legs are forward, meaning his heels are in front of him. He did everything humanly possible while in flight to maximize his landing distance. It sure paid off!

Check out Figure 5.9. Upon landing in the pit, Beamon's body was scrunched up with his center of mass within a few centimeters horizontal distance of his heels' location. Given that his heels and center of mass were not at exactly the same horizontal distance when he landed, my final point in Figure 5.4 may be in error by something on the order of a couple of centimeters. Since I had to estimate the drop in Beamon's center of mass, it is not worth correcting that figure.

Figure 5.9. Beamon touches down well beyond where any other human had reached before. Note that his center of mass is appreciably below the level where it would have been at the launch. (© Bettmann/CORBIS)

Did Mexico City Help Beamon?

We have considered Beamon's jump only as if it had taken place in a vacuum. What about air resistance? As his arms, legs, and torso are rotating, the air is exerting external torque on Beamon. The so-called retarding torque from the air is essentially canceling itself out (though not *exactly*). The reason is that while the arms and torso move, air resisting their motion creates a small torque to the left in Figure 5.8. Likewise, while the legs move forward, the air that resists their motion creates a small torque to the right. The net torque from the air must have been small, because Beamon was not noticeably rotated about his center of mass when he landed in the pit.

What about air drag while Beamon was in flight? I have argued that that force is small compared to the gravitational force. We know from equation (4.2) that a drag force from the air acted on Beamon while in flight. That force will reduce the distance shown in Figure 5.4, meaning of course that I need to increase my launch speed so that Beamon still lands with a record jump of 8.90 m. Then again, how much error did I make?

Beamon weighed about 160 pounds (about 712 newtons) at the time of his jump. From published estimates,[15] $A \sim 0.5\,m^2$ and $C_D \sim 0.9$. Mexico City has an elevation of about 7400 feet (nearly 2300 meters), and it turns out that the air density there is indeed about three-quarters of its value at sea level. Take $\rho \sim 0.9\,kg/m^3$. With a launch speed of $v \sim 10\,m/s$, I use equation (4.2) to estimate a maximum drag force on Beamon of

$$F_D \simeq \tfrac{1}{2}\,(0.9)\,(0.9\,kg/m^3)(0.50\,m^2)\,(10\,m/s)^2,$$

$$\Rightarrow \quad F_D \simeq 20\,\text{newtons} \simeq 4.5\,\text{pounds}. \tag{5.9}$$

Four and a half pounds of air force is a noticeable amount during a long jump. The air drag on Beamon, however, represents at best 3% of his gravitational force (weight). As Beamon soars through the air and rotates his body, his cross-sectional area drops from its running position value. I honestly have no idea what either the drag coefficient or the cross-sectional area is during Beamon's complicated flight. The best I can do, four decades later, is to make a reasonable estimate of the product of $C_D\,A$, much as I did in the last chapter with Lance Armstrong and his bicycle.

If I use the values of C_D and A used in equation (5.9) for the entire flight, and I use the same launch speed and angle that I used to create Figure 5.4, I find that Beamon's center of mass lands at about 8.76 m (roughly 28 ft 9 in). Beamon's heels are ahead of his center of mass; that could make up for the loss of the 14 cm (about $5\tfrac{1}{2}$ in) due to air resistance. Reducing C_D and A will increase the flight distance.

For the model I developed here, Beamon's launch angle and launch speed are roughly what I used to create Figure 5.4. The reduced distance from air resistance is approximately made up by the fact that Beamon's heels are ahead of his center of mass. Remember that I made Figure 5.4 under the assumption that Beamon's center of mass dropped half a meter in the course of the jump. That was my best guess after watching replays of Beamon's jump. Without better camera angles and known standards

[15] See "Effect of wind and altitude on record performance in foot races, pole vault, and long jump," Cliff Frohlich, *American Journal of Physics* 53, 726–30 (1985).

of length in the video of the jump, it would be hard for me to do much better.

In the aftermath of Beamon's miraculous jump, skeptics questioned the favorable jumping conditions Beamon enjoyed. Mexico City's rarefied air is better for jumping in than the air at sea level. Some complained that Beamon had the maximum allowable 2 m/s (about 4.5 mph) of wind at his back. Beamon's jump took place right before it began to rain. His chief competitors thus had to deal with more of the elements. Is there any validity to these concerns?

Let's start with Mexico City. If I use the same C_D and A that I used in equation (5.9), the same launch speed and angle used for Figure 5.4, and the sea-level value of the air density, I find that Beamon's center of mass would have finished at about 8.72 m (roughly 28 ft $7\frac{1}{4}$ in). Mexico City thus helped Beamon by about 18 cm (or about $7\frac{1}{4}$ in). Given that the East German Klaus Beer earned the silver medal with his personal best jump of 8.19 m (about 26 ft $10\frac{1}{2}$ in), Beamon could have jumped anywhere on Earth and he *still* would have crushed his competition! By the way, if $g = 9.80$ m/s^2 at sea level, its value in Mexico City is only about 0.07% less. Being farther from the Earth's center certainly did not give Beamon any added advantage.

What about the 2 m/s wind at Beamon's back? When I do the calculation, I find that the wind probably added about 5 cm (about 2 in) to Beamon's jump. Again, his competition was so far behind him that it was not the wind that gave Beamon the gold. The wind did contribute to the world record, but Beamon did jump under *legal* conditions, albeit the extremes of those conditions. The wind also helped his approach velocity, but again, the gains are not enough to bridge the enormous gap between Beamon's jump and Beer's silver-medal jump.

The bottom line here is that nobody else jumping in Mexico City even came close to Beamon's stupendous jump. Beamon made the perfect approach, hit the take-off board with but a couple of centimeters to spare, and executed perfect technique while in the air. He made one more, almost perfunctory, jump after his record-setter. He jumped 8.04 m (26 ft $4\frac{1}{2}$ in), though I doubt he was trying as hard as he could. For the most part, Beamon's competition gave up after he eclipsed 29 feet. Beamon never reached the 29-ft mark again; he never reached even 27 ft again.

Some leg issues in the early 1970s contributed to Beamon's less than stellar efforts.[16]

For all the critics of Beamon's jump, nobody broke his record until Mike Powell jumped 8.95 m (29 ft $4\frac{1}{4}$ in) in Tokyo on August 30, 1991.[17] That means that nearly 23 years went by. Imagine all the people who long-jumped during those years. Imagine all of the legal jumps in all the various forms of weather, altitude, etc. that were measured and recorded. Nobody could touch Beamon for almost a quarter century. Imagine all the improvements in training, technique, and equipment that would have come along in all those years. No one ever advanced the record by anything like the amount Beamon did. His leap was truly Beamonesque.

Let's now move on to some more exciting Olympic feats, and put our newfound knowledge of angular momentum to good use!

[16] For a wonderful account of Beamon's early life and Olympic times, I strongly urge you to find a copy of *The Perfect Jump* (Signet, 1976) by the late, great Dick Schaap. I can dream that I could one day write half as well as Mr. Schaap.

[17] Powell had a slight tailwind during his jump. At the same event and on the same day, Carl Lewis jumped 8.87 m (29 ft $1\frac{1}{4}$ in) with a headwind about as strong as Powell's tailwind. If only Lewis had jumped when Powell did!

6

Amazing Spins in and around All Kinds of Water

Rotations in Water Sports

Constraints That Liberate

Ding! Have you ever had a bell go off in your head when some idea finally took hold? Did you feel an adrenaline rush and perhaps a few goose bumps? These are what I felt as a university student when I studied *angular momentum conservation* and saw a demonstration like the one I showed you in Figures 5.6 and 5.7. My mind was racing, producing dozens of images of situations where I might apply my newfound understanding of the universe. I thought of all the stuff that turns, rotates, revolves, spins, twists, gyrates, and rolls. Could angular momentum conservation help me understand how all these things work? Sure it could. I had to be careful, though, in how I applied the laws of physics. A football twirled on the grass in the end zone by a player who has just scored a touchdown spends a few seconds spinning before it finally topples over. Angular momentum conservation, with suitable approximations, helps me understand why the football, dancing on its nose, exhibits such remarkable stability. As the ball begins to topple, it becomes apparent that it has lost some of its initial angular momentum. I had to think about external torques acting on the football. But that's okay, because such thinking made the phenomenon more interesting for me.

You may have had the same epiphany I had when I saw the demonstration shown in Figures 5.6 and 5.7. The first image in my head after seeing that demonstration was of an ice skater in her routine's final spin.

The fact that laws of physics put constraints on what the skater could do liberated my mind. I no longer saw the blur of a spinning skater as a mysterious marvel. Suddenly, my passion for sports merged with my burgeoning passion for physics. Physics became "cool" for me, because the same laws that explained the motion of galaxies could explain the beauty of Katarina Witt on the ice, and the majesty of Greg Louganis soaring through the air above a pool.

Poetry on Ice

I once heard it said that if mathematics is the language of the universe, then surely physics is the poetry. In the sporting world, that same sentiment could apply equally to Katarina Witt. She dominated the world of skating in the 1980s like no one else. In that decade, Witt won six straight European Championships, from 1983 to 1988. She also won the World Championships four times (in '84, '85, '87, and '88; a second-place finish in '86 kept her from winning five in a row). Most people probably know of Katarina Witt from her two Olympic gold medals. Representing East Germany, she won gold in both the 1984 Winter Olympics in Sarajevo and the 1988 Winter Olympics in Calgary. Witt dominated her sport with stunning athleticism, and her routines were replete with grace. She was truly poetry on ice.

My favorite Katarina Witt moment came in the 1988 Winter Olympics. Witt was trying to become just the second woman to repeat as Olympic figure skating champion. The incomparable Sonja Henie had taken home three straight golds for Norway back in 1928, 1932, and 1936. Standing in Witt's way were a bevy of talented skaters, most notably the American Debra Thomas and the Canadian Elizabeth Manley. It was Thomas who had claimed the 1986 title at the World Championships, handing Witt her only defeat in a five-year period. After the compulsory figures and the short program, Witt and Thomas were neck-and-neck heading into the long program.[1] The battle between Witt and Thomas was especially

---

[1] The compulsory figures (sometimes called "school figures") portion of a skating program required skaters to create certain "figures" in the ice with their skates. Judges would look at such things as the quality of roundness in a loop. Up until the 1968

thrilling because each skater chose as musical accompaniment a portion of Georges Bizet's rousing opera *Carmen*. The "Battle of the Carmens" would decide the gold!

That Katarina Witt skated before Debra Thomas undoubtedly meant that Thomas would feel added pressure if Witt skated a great long program. Though her routine was not as technically challenging as it could have been, Witt's emotional performance was flawless. Thomas had to know that she needed to perform at an extremely high level if she was going to capture the gold. She chose a challenging routine that included five triple jumps. Perfect execution of all jumps would probably have been enough to push Thomas ahead of Witt. Unfortunately for Thomas, one of her early landings was not perfect. Later in the routine, surely influenced by the pressure to maintain perfection, Thomas suffered another poor landing, needing the assistance of her hand on the ice to maintain stability. Thomas had not only lost the gold, but she lost the silver to Elizabeth Manley; Thomas earned the bronze medal. Manley actually outscored Witt in the long program, but Witt had fared better in the compulsory figures and the short program. The gold was Witt's, and she cemented her legacy as one of the greatest skaters of all time.

What's So Hard about Ice?

We human beings have been traipsing about the Earth in our current form for a couple of hundred thousand years. In all that time, we have relied on water for our existence and our survival. Water has contributed to such forms of enjoyment as swimming pools and skating rinks, and it has had a hand in catastrophe in such forms as icy roads, floods, and avalanches. During the past several hundred years, through the pursuit of sophisticated scientific investigations, we have learned a great deal about water.

Olympics, the compulsory figures had counted 60% of a skater's total score. When Witt won gold in 1984 and 1988, the compulsory figures counted 30% of the total score. The short program, accounting for 20% of the score, takes two minutes and requires skaters to execute certain maneuvers, thus making possible specific comparisons among the skaters. The long program takes about five minutes to complete and counted for 50% of the skater's total score when Witt was winning gold. As the 1980s came to a close, the compulsory figures portion was eliminated.

There are, however, properties of water that still puzzle us today. Before we apply our newfound knowledge of angular momentum to Katarina Witt's skating, let's first examine the wonder of the surface that allows for all the majesty we see in a skater's routine.

Though most of us commonly think of "water" as a liquid, steam and ice are certainly water, too. Steam and ice represent different phases of water. More than for any other molecule, people know the chemical composition of water, H_2O, or "Āch - tü - ō," as one might pronounce it. It rolls readily off the tongue of many, even if they do not understand what the symbols mean. A water molecule consists of two hydrogen atoms and one oxygen atom. A hydrogen atom is made up of a proton and an orbiting electron.[2] In an oxygen atom there are eight protons and eight neutrons in the nucleus, with eight orbiting electrons. Since protons and neutrons are more than 1800 times more massive than electrons, the majority of a water molecule's mass is contained in the nuclei of the three atoms comprising it. Oxygen, in fact, makes up roughly 89% of the mass of a water molecule. A water molecule has a V-shape like that of the ball-and-stick model shown in Figure 6.1. Oxygen is highly electronegative, meaning it attracts nearby electrons. The oxygen atom in Figure 6.1 can thus be thought of as having a slight negative charge. That negative charge attracts hydrogen atoms in nearby water molecules. Neighboring water molecules then form what are called *hydrogen bonds*. These bonds are relatively strong bonds that fall under the description of dipole-dipole interactions. A *dipole* is a separation of positive and negative charge. Because atoms can be polarized (forming a dipole), some neutral objects can be electrically attracted to charged objects. See Figure 6.2 for a toy model of how that works. The electric force is much stronger than the gravitational force, and all it takes for a strong bond is for one sign of an atom's charge to be ever so closer to another atom than the opposite sign of charge. There are about 1.5×10^{23} water molecules in a teaspoon of water, each one bonded to neighboring molecules by hydrogen bonds.

What happens, then, to the nearly 2 trillion trillion water molecules in a teaspoon of water when the temperature drops low enough to form ice?

---

[2] Different *isotopes* of atoms are made by adding one or more neutrons. So-called heavy water replaces normal hydrogen with deuterium, an isotope of hydrogen with a neutron in the nucleus.

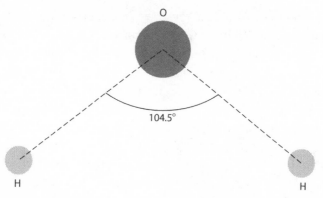

Figure 6.1. A ball-and-stick model of a water molecule. The separation between a hydrogen atom and the oxygen atom is just under 1 Å (recall that an angstrom is 10^{-10} m).

Figure 6.2. Dipole-dipole interaction. Two neutral atoms become polarized; their opposite charges attract. Two atoms (or molecules) with no overall charge thus attract each other.

Ice is water in a solid state. Unlike most substances, water is *less* dense in its solid form than in its liquid form. Have you ever thought about that while enjoying a refreshing glass of ice water? Ice floats in liquid water; icebergs float in the ocean. As liquid water is cooled, the hydrogen bonds formed are such that the water molecules form hexagonal shapes. That hexagonal shape partially explains the shape of snowflakes (temperature and water content in the air also influence snowflake shape). Compared to water in its liquid phase, the hexagonal packing of ice crystals is not as dense. At normal atmospheric pressure and at a temperature of about 4°C (roughly 39°F), water has its greatest density—about 1.00 g/cm³ (one gram per cubic centimeter). When water freezes at 0°C (32°F), its density is about 0.917 g/cm³, about 8.3% less than water at 4°C. This fact explains why aquatic life can survive in ponds and lakes that freeze over in the winter. As water cools to 4°C in a pond, it sinks, because it is more dense than water at other temperatures. But water cooler than 4°C is less dense, and therefore stays near the surface. As the temperature drops further, water at the surface freezes. Life below survives in relatively comfy 4°C liquid water.

What do all these facts about water have to do with Katarina Witt's skating? We are trying to understand why Witt and other skaters can so easily glide on the ice. Why is there so little friction between the ice and the skates? I've given you a qualitative idea of what happens to water as it is cooled. Changing pressure can also affect the melting point of water (i.e. the temperature at which water changes phase from solid to liquid). Because of the peculiar density property of water—ice is less dense than liquid water, the melting point of water actually *decreases* as the external pressure increases. To see why, think about what happens if you exert pressure on an object. The more pressure you apply, the more you reduce the object's volume, meaning that its density increases. For ice at, say, 0°C, increasing its density by increasing the external pressure on it means that the ice can change to a liquid. It was once commonly believed that the force per unit area—pressure—that a skate exerts on the ice caused a very thin layer of ice to change to liquid water and thus make ice slippery.

This explanation, about why there is little friction between ice and skates, sounds reasonable, especially since water possesses the rare property of having its melting point drop as external pressure increases. The reason I used the term "once commonly believed" is because experiments performed in the past few years tell a different story.[3] It turns out that the pressure created by a skater on skates is not nearly great enough to melt a layer of ice sufficient for low-friction gliding.

The answer to the slippery question lies possibly in one of two explanations, or a combination of the two. The first explanation relies on frictional heating of the ice by the skating blade. When two surfaces rub against each other, bonds are continuously formed and broken between molecules in the two surfaces. There is thus a resistance to the motion— *kinetic friction*. When molecules form and break bonds, they vibrate more than normal. The additional vibration is a sign that energy has been added to the molecules. The associated temperature increase gives rise to the

---

[3] What I know about the slippery properties of ice began when I read a fine article: "Why Is Ice Slippery?" by Robert Rosenberg, December 2005 issue of *Physics Today*, pages 50–55. A less technical article dealing with the same topic is: "Explaining Ice: The Answers Are Slippery" by Kenneth Chang, February 21, 2006, issue of *The New York Times*.

possibility of surface melting. I use the word "possibility" because if the ice is much colder than the melting temperature, the water layer produced via frictional heating will most likely be too thin to allow for smoother skating. According to Rosenberg's article, the "optimum temperature for figure skating is −5.5°C," probably close enough to the melting temperature for frictional heating to be a candidate for explaining why ice is slippery.

A second explanation has to do with the surface properties of ice. An object's surface often behaves differently from the object's interior, or bulk. An easy way to think about why this is so is to imagine sitting on a molecule in the interior of a chunk of ice. What you see to the left looks the same as what you see to the right. The same goes for up and down, and for front and back. In other words, there is "translational symmetry" in the bulk. As another thought experiment, imagine a microscopic you hopping from one water molecule to the next, within the bulk; everything would look the same. At the surface, however, the symmetry is broken. You look down and you see ice; you look up and you see an ice skate over your head. The water molecules at the surface of ice behave differently from those in the bulk, because the surface molecules feel only bonds from below. They are not held in place as tightly as those in the bulk because there are no water molecules above them to bond with. It is thus believed, with experimental evidence to back the claim, that water molecules at the surface vibrate more than those in the bulk and therefore remain in a quasi-liquid state. This liquid-like layer allows a skate to glide across the ice.

Whether friction, an already present liquid-like layer, or a combination of the two explains why ice is slippery, the idea of pressure melting in ice skating has been all but abandoned. What is true is that the graceful strides of Katarina Witt on the ice are not as easily explained as one might have thought. A great deal of physics and chemistry is needed!

Back to Spins

Let's now put to good use the angular momentum knowledge we acquired in chapter 5. Figure 6.3 shows Katarina Witt in a mid-air spin during the short program of the 1984 Olympics. Think about equation (5.5) and note the configuration of Witt's arms and legs. (I tried much less gracefully to

Figure 6.3. Katarina Witt in a tight aerial spin at the 1984 Winter Olympics. Note how her arms and legs are close to her vertical axis. (© Bettmann/CORBIS)

do something similar in Figure 5.7.) A fast spin is achieved by pulling one's mass as close to the axis of rotation as possible, thus reducing one's moment of inertia.

What does Katarina Witt have to do to spin so fast? According to equation (5.8), angular momentum conservation dictates that over a short period of time, the product of her moment of inertia and her angular velocity is roughly a constant. Witt can thus spin fast by reducing her moment of inertia, i.e. she pulls her arms and legs toward her center. Spinning fast makes it tough for one to control oneself; skaters are not spinning during their entire routines. What Witt does is to begin spinning slowly, with a large moment of inertia. Figure 6.4 shows Katarina Witt at the 1994 Olympics in Lillehammer, Norway.[4] Note again the configuration of Witt's arms and legs: she has moved as much mass as she can from her vertical axis of rotation. Since she has to stand on at least one leg, she moves both arms and one leg far from her center, raising her moment of inertia. Because the product of moment of inertia and angular velocity is roughly constant over a short period of time, Witt can double her spin rate if she can reduce her moment of inertia by a factor of two. (I tried something similar in Figure 5.6.) If you have access to a low-friction swivel chair or platform, give it a try. The feeling you get while pulling your arms and legs in and thus getting your angular speed revved up is thrilling.

We know how to explain Katarina Witt's fast spins. What is happening with energy during that time when her spin rate increases? You may have heard of *energy conservation* before. The idea behind that powerful conservation law is that the total energy in a closed system remains constant in time. The "closed system" concept is omnipresent in physics. We apply energy conservation to systems that cannot exchange energy with their external environments. If Katarina Witt shoves off the ice and does nothing else while gliding on the ice, she will eventually come to rest, because she loses her energy of motion, owing to the frictional interactions between her skates and the ice and between the air and her body. If our model system consists of only Katarina Witt, total energy is not conserved, since energy leaks out of the model system. If, by contrast, our

[4] Witt finished seventh in the 1994 Winter Olympics. Her skating focused more on her urging for peace in Sarajevo, the site of her 1984 gold medal and, in 1994, a city under military siege.

Figure 6.4. Katarina Witt personified grace at the 1994 Winter Olympics. Note how her arms and back leg are far from her vertical axis. (© Wally McNamee/CORBIS)

model system is made up of Witt and her surroundings, specifically the ice and the air, total energy is conserved.

But we need to understand in a qualitative way what is happening with energy while Katarina Witt goes into a fast spin. I will thus make some approximations. Assume that the friction between her skates and the ice is sufficiently small that on the time scale of her spin, only a small amount of energy is lost. The frictional losses with the air will also be ignored. I assume further that Witt's center of mass does not move, so that I may ignore translational motion. We can do a fast calculation without any numbers to get a feeling for what happens to the energy. First, we need an expression for the energy of motion associated with spinning. This *rotational kinetic energy* about a chosen axis is defined as

$$KE^{\text{rot}} = \tfrac{1}{2} I \omega^2, \tag{6.1}$$

where the "rot" superscript means *rotational*.[5] Suppose that in bringing in her arms and leg for a fast spin, Katarina Witt is able to reduce her moment of inertia by a factor of two. Conservation of angular momentum tells us that her angular speed must increase by a factor of two. We can write those statements mathematically as

$$I_f = \tfrac{1}{2} I_i \tag{6.2}$$

and

$$\omega_f = 2 \omega_i, \tag{6.3}$$

where "i" means *initial* and "f" means *final*. Witt's initial rotational kinetic energy is

$$KE_i^{\text{rot}} = \tfrac{1}{2} I_i \omega_i^2, \tag{6.4}$$

whereas her final kinetic energy is

$$KE_f^{\text{rot}} = \tfrac{1}{2} I_f \omega_f^2. \tag{6.5}$$

---

[5] As I mentioned in footnote 6 in chapter 5, the moment of inertia is really a *tensor*. Equation (6.1) is thus a simplified version of a more formal equation. For our purposes here, consider a vertical axis of rotation used to determine the moment of inertia and the angular velocity. We certainly do not need tensors to gain a qualitative feel for what's going on.

Now plug equations (6.2) and (6.3) into equation (6.5), and use equation (6.4) to get

$$KE_f^{rot} = \tfrac{1}{2}\left(\tfrac{1}{2} I_i\right)(2\,\omega_i)^2 = 2 \cdot \tfrac{1}{2} I_i\,\omega_i^2,$$

$$\Rightarrow KE_f^{rot} = 2\,KE_i^{rot}. \tag{6.6}$$

Wait a minute! What happened here? We were assuming that Katarina Witt was the only object in our closed system. How could Witt's rotational energy have *doubled*[6] between the time when her arms and leg were fully extended and the time when her arms and leg were pulled tightly against her body? From where did the energy come?

Witt, then, must do mechanical work on her arms and outstretched leg as she pulls them closer to her rotation axis. Mechanical work is the product of a force with the distance an object moves along that line of force. The energy source for that mechanical work comes from a conversion of chemical energy in Witt's body to mechanical energy. Analyzing such a system can be tricky, because it is not always obvious where energy can flow. We certainly do not "see" the chemical energy being converted to mechanical energy. We do, however, see the *effect* of such a transfer, in the sense that we see our arms and legs moving.

We will examine energy processes in the body in much greater detail in chapter 9. For now, let's estimate how much energy is gained by pulling one's arms and leg in while setting up a fast spin. Using equations (6.5) and (6.6), the change in rotational kinetic energy may be written as

$$\Delta KE = KE_f^{rot} - KE_i^{rot} = \tfrac{1}{2} KE_f^{rot},$$

$$\Rightarrow \Delta KE = \tfrac{1}{2}\left(\tfrac{1}{2} I_f\,\omega_f^2\right). \tag{6.7}$$

We thus need to know Katarina Witt's moment of inertia while spinning fast, and we need to know her rotational speed. Since I am interested only in a back-of-the-envelope estimate of Witt's rotational kinetic energy increase, I will model Witt as a uniformly dense cylinder that rotates about its longitudinal axis. Katarina Witt is clearly not a cylinder, but we will be able to produce a reasonable order-of-magnitude estimate of her moment of inertia while in the fast spin configuration.

---

[6] One can show that the ratio of final rotational kinetic energy to initial rotational kinetic energy is ω_f/ω_i, which is the same as the ratio I_i/I_f. With those expressions, you can play with numbers beyond my factor-of-two example.

It turns out that for a uniform cylinder of mass M and radius R, its moment of inertia about its longitudinal axis is $MR^2/2$. I estimate Witt's mass to be about 55 kg and her radius while in a fast spin to be roughly 12 cm. Thus,

$$I_f \sim \tfrac{1}{2} MR^2 \sim \tfrac{1}{2} (55\,\text{kg})(0.12\,\text{m})^2 = 0.396\,\text{kg}\cdot\text{m}^2 \sim 0.4\,\text{kg}\cdot\text{m}^2, \quad (6.8)$$

where I have kept only one significant digit for my rough estimate.

For the final rotational speed, I estimate Witt's final spin rate to be around three revolutions per second. Recalling footnote 10 in chapter 5, we need the rotational speed in radians per second. Thus,

$$\omega_f \sim \left(\frac{3\,\text{rev}}{\text{s}}\right)\left(\frac{2\pi\,\text{rad}}{1\,\text{rev}}\right) \simeq 18.8\,\frac{\text{rad}}{\text{s}} \sim 20\,\frac{\text{rad}}{\text{s}}, \quad (6.9)$$

where I have once again kept only one significant digit.

Now, plug equations (6.8) and (6.9) into equation (6.7) and get

$$\Delta KE \sim \tfrac{1}{2}\left[\tfrac{1}{2}(0.4\,\text{kg}\cdot\text{m}^2)\left(20\,\frac{\text{rad}}{\text{s}}\right)^2\right],$$

$$\Rightarrow \Delta KE \sim 40\,\text{J} \sim 0.01\,\text{kcal}, \quad (6.10)$$

where I have used the conversion between joules and nutritional calories (kilocalories) given in footnote 10 in chapter 4. The laws of thermodynamics tell us that no engine can convert all of its input energy into useful work. There is always some waste heat. My estimate for Katarina Witt's increase in rotational kinetic energy is therefore an underestimate of the number of nutritional calories she needs to burn to go from a slow spin to a fast spin. The main point is that there are at best a few *hundredths* of a nutritional calorie burned when Witt pulls her arms and leg in close to her body. That is not much energy when compared to the roughly 2000 nutritional calories that a person consumes in a day.

Majesty above the Water

During the same two Olympic years in which Katarina Witt was establishing herself as the premier figure skater of the decade, Greg Louganis was making the case that he might be the greatest diver of all time. Diving for the United States, Louganis brought home gold in both the 3 m spring-board and the 10 m platform events in both the 1984 Summer Olympics in Los Angeles and the 1988 Summer Olympics in Seoul, South Korea. By earning back-to-back Olympic golds in the two most prestigious diving events, Louganis accomplished what only the great Patricia McCormick had achieved.[7]

We will never know the full extent of what the U.S. boycott of the 1980 Summer Olympics in Moscow may have cost American athletes. My guess is that Louganis would have enjoyed double gold for the *third* straight time in Seoul, instead of just his repeat in 1988. Why do I believe that? After winning the silver medal at only 16 years of age in the 10 m platform at the 1976 Summer Olympics in Montreal, Quebec, Canada, Louganis won the World Championship 10 m platform title in Berlin, West Germany, in 1978. He also won titles in the 10 m platform and 3 m springboard events at the 1982 (in Guayaquil, Ecuador) and 1986 (in Madrid, Spain) World Championships. Louganis also won gold medals in the 10 m platform and 3 m springboard events at the 1979 Pan Am Games in San Juan, Puerto Rico. He repeated his double gold at the 1983 Pan Am Games in Caracas, Venezuela. A third straight set of two gold medals hung from Louganis's neck at the 1987 Pan Am Games in Indianapolis. With all of those gold medals that Louganis earned before and after the 1980 Summer Olympics, my guess is that he would have had two more Olympic golds in his trophy case had the United States participated in the Moscow Olympics.

My favorite Greg Louganis diving moment came in the 1988 Seoul Olympics. During one of his preliminary 3 m springboard dives, Louganis opted for a reverse $2\frac{1}{2}$ somersault pike dive. Unfortunately, he failed to completely clear the diving board and hit his head on the end of the board. After getting stitched up, he came back and won the 3 m springboard gold. With stitched head, Louganis then completed the double gold repeat with a win in the 10 m platform.

In the sections that follow, I will discuss Louganis and springboard div-ing. Platform diving has its own intricacies; looking at springboard diving

---

[7] Diving for the United States, McCormick earned one of her two golds in the 1952 Summer Olympics in Helsinki, Finland, and the other in the 1956 Summer Olympics in Melbourne, Australia. McCormick's victory margin in the 1956 3 m springboard is the greatest in Olympic diving history (at least as of the time of this writing).

only, however, will allow me to cover most of the physics we need to understand diving.

Springy Boards

You know that a discussion of angular momentum is coming. Let's first, however, get Louganis into the air before applying angular momentum conservation to all the twists and somersaults he executed. How easy does it seem to you when we see Louganis leaping off the end of a 3 m springboard and into the pool? You may never have jumped from a 3 m springboard, but you might have jumped off of a 1 m springboard at a public pool. My guess is that if you tried to mimic what you have seen professional divers do, namely take a few steps to the end of the board and then hop high into the air, you probably noticed that the board's vibration as you got ready to hop off made you feel a little uneasy. That's how I felt the few dozen or so times I tried to dive off of a springboard. In my younger days, before I had any idea what a diver needs to do before jumping from a springboard, I would strut along the board thinking I was about to leap way into the air and execute a Louganis-like dive. Instead, the vibrating board made me feel like I was about to fall off. I thus reached the end of the board, stopped, and stepped off with an inglorious "dive." After watching how divers approached their takeoff, I started to wonder if there wasn't something to all of the motions the divers went through on the board. It turns out that all those motions are critically important to good dives, and physics can tell us why.

Figure 6.5 shows a rough sketch of a springboard used for diving. According to the USA Diving Rules & Code, Section 101.2, the springboard "shall be approximately 20 inches wide and 16 feet long" (50.8 cm wide and about 4.88 m long). "A mechanically adjustable fulcrum of a type readily adjustable between dives" (see further discussion in Section 101.2 of the rules) is placed typically about one-third of the way along the board from its fixed end. A standard technique[8] for approaching the

---

[8] There are several books on springboard diving. A particularly good one is Charles Batterman's *The Techniques of Springboard Diving* (MIT Press, 1968).

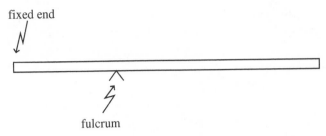

fixed end

fulcrum

Figure 6.5. A sketch of a springboard used for diving. The fulcrum is roughly one-third of the way down the board from the fixed end.

end of the board is to begin with three balanced and rhythmic steps. The third step takes the driver about 1 m past the fulcrum. These rhythmic steps have initiated small oscillations in the board, and the third step leads immediately into what is called the "hurdle," made hopefully with the help of the upward motion of the oscillating board.

To initiate the hurdle, the diver pushes the board down with one foot while pulling the other foot and both arms upward. Newton's third law says that there must be a reaction to the upward force needed to elevate the arms and leg. This downward reaction force further assists the diver in getting the board to depress. The depressed board stores elastic potential energy in the same way the ruler in Figure 1.2 does. The microscopic stretching of molecular bonds in the board is analogous to stretching a spring. You know that if you stretch a spring and let go of it, the spring moves back toward its original location. The diving board does the same thing. The board's reaction pushes the diver upward during the hurdle. Because the diver was moving forward at the time of the hurdle, there is a forward component to the approach velocity while the diver is in the air getting ready for the big jump off the end of the board. That forward motion helps the diver clear the end of the board when the final jump takes place.

Typically, the hurdle is timed such that the springboard oscillates $2\frac{1}{2}$ times before the diver lands on the end of the board. That means the diver lands on the end of the board while the board is moving downward with its greatest speed. With the board moving downward, the diver seeks to maximize the board's depression. A deep depression means a considerable amount of stored elastic potential energy. Much of that stored energy is given to the diver to get him or her as high off the board as possible.

Think about how conservation of energy plays a role in getting a diver propelled into the air. The diver transfers some energy to the board during the approach. Though there are times when transferring energy during walking may be disadvantageous,[9] the diver in our analysis benefits because some of the board's energy is transferred back to the diver during the hurdle. When the diver leaves the end of the board at takeoff, some of the board's stored potential energy is transferred to the diver, thus allowing the diver to leave the board with a speed greater than what could be obtained with no board oscillations. The various energy transfers from the diver to the board or vice versa are never complete. For example, when you hear the board make its distinctive noise as the diver leaves it, your ear absorbs energy in the form of sound waves. We encountered this form of energy transfer back in chapter 1 when we discussed the crack of a baseball bat against a ball.

Louganis and Falling Cats

What is so special about getting help from the springboard? Why is a diver so concerned about gaining as much height as possible for the 3 m springboard event? Think about the kinematics. Once Louganis leaves the board, his center-of-mass motion is completely determined by the external forces acting on him. Though some air resistance is involved, the gravitational force is by far the greatest force Louganis feels while in the air. If we assume, for simplicity, that gravity is the only force on Louganis, the kinematic equations we used back in chapter 3 describe the trajectory of his center of mass.

---

[9] There are times when *resonance*, i.e. the large amplitude oscillations that arise when input energy is given to a system in just the right way, is good. When the voice of a great tenor like the late Luciano Pavarotti resonated, the result was an enjoyable and immensely pleasing sound. There are other times, however, when resonance is not so desirable. The Angers Bridge in France collapsed in 1850 when French soldiers marching in lockstep caused large vibrations in the bridge. The bridge fell and an estimated 226 soldiers perished. In the century and a half since then, soldiers worldwide have been instructed to break step when marching across a bridge, so as to avoid resonating the bridge.

I watched video of several of Louganis's Olympic 3 m springboard dives. In a typical dive, he leaves the springboard and reaches the water in a time of about 1.7 seconds. According to the United States Olympic Committee's web site profile of Greg Louganis,[10] his own height is 5 feet 9 inches (about 1.75 m). For the dive, then, I assumed his center of mass to be one meter from his feet, leaving about 0.75 meters from his center of mass to the top of his head (torsos are heavier than legs). When Louganis leaves the springboard at a height of 3 m, his center of mass is 4 m above the surface of the pool. My time estimate of 1.7 seconds is the time it takes for his feet to leave the board and his head to reach the water. His hands and forearms are in the water by the time his head gets wet. Since cleaving the water cleanly is critical to a diver's score, I ignore the force from the water on his arms during the short time needed to submerge half his arm length. His center of mass's final height above the water is thus 0.75 m. I further estimated that when his center of mass returned to the same height as the board, i.e., 3 m high, it was about 1 m away from the board. Thus, $x = 1$ m and $y = 3$ m is a point I insist be on the parabola.

Using the kinematic equations from chapter 3 and the assumptions from the previous paragraph, I find that Louganis's launch speed is about 6.46 m/s (roughly 14.4 mph) and his launch angle is around 83.9°. Note that a launch angle much closer to 90° is dangerous, because there may not be enough horizontal velocity to ensure that the diver is clear of the board on the way down. Remember that I assumed Louganis's center of mass to be 1 m from the board; other parts of his body are closer. Figure 6.6 shows the trajectory of Louganis's center of mass during a typical Olympic 3 m springboard dive. About 1.45 s, or roughly 85% of the 1.7 s of flight time, have elapsed by the time his center of mass reaches the height of the board ($y = 3$ m). By the time his center of mass passes the board, Louganis is getting in position for a straight entry into the pool. That nearly second and a half time spent above the diving board is all the time Louganis has to execute all of his spins and twists. You can thus understand why a diver wants to get as much launch speed as possible. The longer in the air, the more complicated the dive that can be attempted. The more complicated the dive, the better chance for high scores from the

[10] See www.usoc.org/26_38190.htm.

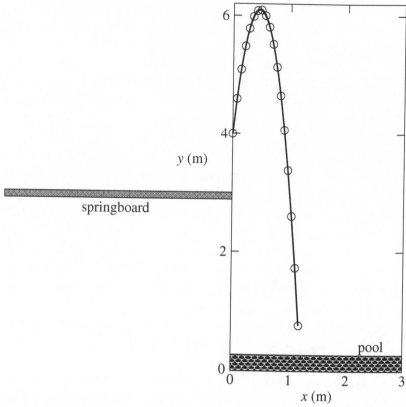

Figure 6.6. Center-of-mass motion of a Greg Louganis 3 m springboard dive. The launch speed is $v_0 \simeq 6.46$ m/s and the launch angle is $\theta \simeq 83.9°$. The open circles represent the location of Louganis's center of mass at 0.1-second intervals.

judges, since competitors can earn more points by performing difficult dives than they can with relatively easy dives.[11]

Let's now investigate how Louganis turned a mere second and a half into Olympic gold. To execute $2\frac{1}{2}$, or even $3\frac{1}{2}$, somersaults, Louganis must spin at a fast rate. Tucked into a tight ball, he will be in the aptly

[11] See Article 7 of the *USA Diving Rules & Code* for information about judging. If you have never seen the scoring rules, you may be surprised by how intricate and detailed they are. As an adolescent watching Louganis attain Olympic glory, I thought naïvely that the judges watched his dive and then settled on scores that simply represented their opinion of the dive's quality. I was completely wrong!

named "tuck" position. If we imagine three complete turns in 1.5 seconds, equation (5.6) then tells us his average angular speed. Three revolutions in 1.5 s means $\omega = 2\,\text{rev/s} \simeq 12.6\,\text{rad/s}$, or you can think of the angular speed as 120 revolutions per minute.

That angular speed is about half the speed of an audio compact disc when read near its outer edge. Many compact disc readers rely on a constant *linear* speed for the portion of the disc they are scanning. Since the outer rim of the disc has greater linear speed than the inner portions (in a given amount of time, a point on the rim travels a greater distance than a point nearer to the disk's center), disc readers must slow a disc down as the scanning moves outward. When scanning near the center of the disk, the angular speed is about $2\frac{1}{2}$ times what it is when the reader scans the disk at the rim. In physics, we relate the linear speed, v, to the angular speed, ω, by $v = r\omega$, where r is the radial distance from the axis of rotation to the point of interest. An angular speed of 120 rpm is pretty fast, and it takes a great deal of athleticism and concentration to be able do what Louganis did.

We know that to enter the water, Louganis wants as little splash as possible. He must arrest all of his fast spinning before he enters the water. How does he do it? Angular momentum conservation! Just as Katarina Witt can move her arms outward to slow her spin, Louganis moves his arms and legs far from his center of mass, so that he may enter the water with little rotation. Figure 6.7 shows Louganis entering the water in the "layout" position with his body as long as possible. The "layout" position has about four times the moment of inertia as the "tuck" position.[12] Angular momentum conservation tells us that if Louganis's moment of inertia quadruples, his angular speed must drop simultaneously by a factor of four. Dropping to half a revolution per second gives Louganis enough control when he enters the water, since, as Figure 6.6 shows us, he spends only a couple of tenths of a second below the board. With an angular speed of about 0.5 rev/s, Louganis rotates only about a tenth of a full turn (36°) in 0.2 seconds. He must come out of the "tuck" position at just the right moment, so that a small rotation in the "layout" position gets his body completely vertical for the entry into the pool.

[12] See "Do springboard divers violate angular momentum conservation?" by Cliff Frohlich, *American Journal of Physics* 47, 583–92 (1979).

Figure 6.7. Greg Louganis enters the pool during the 1984 Olympics. Not much splash, is there? (© Duomo/CORBIS)

What about twists? To makes things more interesting, with twists, check this out from Section 107.4 of the *USA Diving Rules & Code*: ". . . the twisting shall not manifestly be done from the board or platform." That's quite a rule! What that means is that Louganis is not allowed to use the board to create the torque needed to twist. With somersaults, Louganis makes use of relatively small horizontal friction forces to help generate the necessary torque. He can't do that with twists. Once in the air, he has only his body movements to help him. He certainly makes use of his body movements when somersaulting; with twists, there is nothing else to use.

Louganis is analogous to a cat dropped upside down. While I do *not* recommend trying this little demonstration, since it's probably not that much fun for the cat, I *will* tell you that cats have a way of landing right-side up. For both a falling cat and Greg Louganis executing a series of twists, the total angular momentum is conserved. The cat turns over before hitting the ground by first bending at its waist. Louganis does something similar, but he must do more because the beauty of a twist in diving is that the diver's body is completely straight while executing it.

To understand how Louganis twists, check out Figure 6.8. He has thrown his right arm over his head and rotated his left arm below his shoulder and toward his midsection. Why? Upon leaving the diving board, Louganis threw his arms and upper body slightly backward. In addition to keeping his feet from slipping on the board during that maneuver, the board's static friction creates a torque about Louganis's center of mass. Louganis's head rotates away from an observer on the opposite end of the pool, since the board's static friction force is toward the observer. Using the right-hand rule, the observer would conclude that Louganis's angular momentum vector points to the observer's left. Louganis thus begins his springboard dive with a nonzero angular momentum, which cannot be altered while in flight.

We have seen that after leaving the springboard, Louganis immediately throws his right arm over his head and moves his left arm down toward his side. If we were to float in the air and look at Louganis from his front (toward the board), we would see his arms rotating in a clockwise direction. To keep his angular momentum fixed in both magnitude *and* direction, the rest of his body begins to rotate counterclockwise. I use the word "begins" because his body cannot stay in the same plane once

Figure 6.8. Greg Louganis in the middle of a twist during the 1988 Olympics. Note the configuration of his arms. (© Dimitri Iundt/TempSport/CORBIS)

the arm movement starts. Remember that Louganis left the board with a backward rotation. Once his body begins to rotate, it must also twist, to keep his angular momentum vector pointing in the same direction with the same magnitude. The twisting continues while Louganis holds his arms as in Figure 6.8. He arrests his twisting motion by moving his arms back out and bending at the waist. Before you try one of these twisting dives, keep in mind how much strength, training, and sheer athleticism it takes to do what Louganis did. I've watched his $\frac{1}{2}$ somersault, $3\frac{1}{2}$ twist springboard dive from the 1988 Seoul Olympics a few dozen times, and I still cannot believe a human being is capable of doing what he did. By the way, his entry into the water might have knocked a half dozen water droplets into the air!

Help Me, Archimedes!

Appendix B of the USA *Diving Rules & Code* says that the "preferred" water depths for the 3 m springboard and 10 m platform dives are 3.80 m and 5.00 m, respectively. We think of diving as a "pool sport." It is, but except for the entry into the water, a diver's points are determined above the water's surface. Although it is silly to call diving an "air sport," we can think now about what role the water plays in diving. As much as fans and judges love to see a perfect entry into the water, with as little splash as possible, divers use the water for something entirely more vital. They need to stop falling!

In the model trajectory I showed you in Figure 6.6, Louganis's center of mass is moving at a speed of roughly 10 m/s (~22 mph) when his hands hit the surface of the water. If after a springboard dive, Louganis had to fall another 3.80 m through air, his speed would be about 13 m/s (~29 mph). I think we all agree on the fact that nobody wants to dive off a springboard and hit cement at nearly 30 mph! The water in the pool provides a drag force sufficiently large to slow the entering diver down to zero speed. In chapter 4, we employed a model for drag given by equation (4.2). That model equation tells us that the drag force is proportional to the density of the material through which an object moves. When Lance Armstrong was biking through the French Alps, he moved through air with a density of about 1.2 kg/m^3. Louganis is slowed by water with a density of

$1000 \, kg/m^3$, roughly 800 times denser than air. Whereas a cubic meter of air weighs only about 2.6 pounds, a cubic meter of water weighs around a ton! Louganis feels far more drag force from the water than Armstrong feels from the air.

If he were to keep his entry-into-the-water form all the way to the bottom of the pool, Louganis would reach the cement in about a second. What divers often do, however, is alter their cross-sectional area just after they enter the water. They angle their bodies so as to follow a curved arc, and they spread their arms out. Because humans are about 6% to 7% denser than water, we do not float in water. If Louganis were to enter an infinitely deep pool and not change his body's configuration during his descent into the water, he would eventually reach a terminal speed. That would occur when his acceleration drops to zero, meaning, by equation (3.6)—Newton's second law—that the net force on Louganis is zero. That does not mean that there are no forces present. Indeed, there are forces pulling Louganis up and the gravitational force is pulling him down. When the net upward force matches the net downward force, the speed can no longer change. Of course, Louganis never reaches terminal speed because he needs to come up to breathe!

I mentioned upward forces in the preceding paragraph. In addition to an upward drag force, there is another force pushing Louganis up toward the surface. Because Louganis displaces water while in the pool, he feels an upward buoyant force. Any object immersed either partially or fully in any type of fluid feels a buoyant force. Remember that a fluid is a liquid or a gas. A balloon filled with helium "floats" in the air because helium is less dense than air and a helium-filled balloon displaces air. The air is responsible for an upward buoyant force on the balloon.

Buoyancy was first conceived by Archimedes[13] who, when contemplating how to determine if the king's crown was made of gold or was a phony, hopped exuberantly from his bathtub and ran through the streets of Syracuse exclaiming, "Eureka!" ("I have found it!"). *Archimedes' Principle,*

---

[13] Archimedes was a Greek scientist who lived from 282 BC to 212 BC. Among his many accomplishments, he invented a screw that helped get water from the ground and raise water from rivers. During the Roman siege of Syracuse in 212 BC, a soldier tried to take Archimedes from his home. Deep in mathematical thought (according to some accounts), Archimedes refused, and the Roman soldier stabbed him to death. The sword is mightier than the protractor!

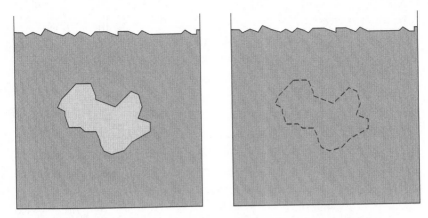

Figure 6.9. Conceptual proof of Archimedes' Principle. On the left is an object submerged in a fluid in a container. On the right, the object has been replaced by a make-believe bag with the exact same size and shape as the original object, but now the space is occupied by the fluid.

as we call it now, states that the buoyant force on an object from a fluid is the same as the weight of the amount of fluid displaced by the object. If Archimedes were to note the water level in his tub before he got in it, and then note the level after he got in, he could determine the volume of water he displaced. The weight of that displaced water is the buoyant force Archimedes would have felt. Even though I have avoided proofs and mathematical rigor in much of this book, I feel I should show you a proof of Archimedes' Principle, because the proof is so simple. Figure 6.9 shows two figures. On the left is an object submerged in a fluid. The object could be a baseball, Greg Louganis, an okapi, or anything else. The fluid is also arbitrary; it could be the air, water, helium, beer, etc. The object feels pressure from the fluid; the associated forces push on the object from all sides, always perpendicular to any small area on the object. On the right side of Figure 6.9, I show you the exact same situation as on the left, except that I have removed the object. There is only fluid in the container. I did leave, however, an imaginary object with the exact same size and shape as the original object. There is nothing there, of course, only the dotted outline of where the object used to be. Instead of the original object occupying the space inside the dotted outline, the surrounding fluid occupies it. With the object removed, the fluid has settled down (fluid has filled its void) and is in static equilibrium. Whatever force from the fluid the object felt must be the same force the fluid contained in our imaginary bag feels. Because the fluid in the make-believe bag feels a downward force from gravity, i.e. its weight, and because the make-believe bag feels no net force, the upward force must equal the weight of the fluid. Working backwards to the object in the container on the left side of Figure 6.9, we see that the upward buoyant force on the object must equal the weight of the displaced fluid. *Quod erat demonstrandum!*

Applications of buoyancy are abundant. People enjoy hot-air balloon rides in which the heated air within the balloon is less dense than the cooler air outside the balloon. Water therapy is used by physical therapists not only because water offers greater resistance than air, but also because the buoyant force reduces the amount of stress on patients' joints. Buoyancy (as in a pool) assists patients with weak muscles, like those who suffer from arthritis or multiple sclerosis, as they exercise their limbs. Evolutionists theorize that large animals like whales grew enormous in the oceans because buoyancy helped support their weight (legs were no longer needed). The internal organs of today's whales can be crushed under their own weight if a whale is removed from the water. Buoyant forces also explain why ice floats. We have already seen that ice is less dense than liquid water. One can show that the fraction of a floating object beneath the surface of the fluid in which the object floats is $\rho_{object}/\rho_{fluid}$ (remember that ρ is the mass density). About $\frac{9}{10}$ of ice is beneath the surface of water, a fact to which sailors pay special heed. Water at 3.98°C has a density of 1.000 g/cm^3. Seawater is slightly more dense; at 15°C its mass density is 1.025 g/cm^3. These numbers do not alter the fact that roughly $\frac{9}{10}$ of an iceberg is beneath the ocean's surface. Lastly, on my limited survey of applications of buoyancy, I have it on my "to do" list to visit the Dead Sea one day. About nine times saltier than normal seawater, and thus more dense than seawater, the Dead Sea permits visitors to float within it, effortlessly.

Going back to Greg Louganis entering the water, he benefits from the water's drag force and buoyancy, which work against the gravitational force. When he gets back to the water's surface, Louganis is further helped by buoyancy. His arms and legs need only apply a modest downward force to the water to keep him afloat. Treading water is certainly easier than trying to tread air! If we could do that, we could fly.

A Few More Tidbits

Energy conservation is a profound idea, but it is relatively simple to state. The total amount of energy in a closed system is fixed. The various forms of energy—kinetic, elastic, gravitational, potential, electrical, etc.—may change. When all forms of energy are added up, the total remains fixed in time. Imagine Greg Louganis entering the water as in Figure 6.7. Moving at a speed of about 10 m/s, he has a kinetic energy of roughly 3750 J. To obtain that calculation, I assume a mass of 75 kg (weight around 165 lbs.) for Louganis. Whether or not that is an accurate estimate of his mass in the 1980s does not change the qualitative discussion in this chapter. When he gets back up to the surface, his kinetic energy is nowhere near 3750 J. His gravitational potential energy is the same, since I am now looking at Louganis at the water's surface. Where did all that kinetic energy go? A small amount went into the sound wave that we hear as a "splash." Most of the energy is lost to the water. As Louganis descends into the pool, his body does work on the water he pushes. By increasing the agitation among the water molecules, Louganis causes a slight temperature increase in the water. The specific heat of water, which measures the amount of energy that must be added to a given mass of water to raise its temperature by a certain amount, is 4186 J/kg·°C. That means that it takes 4186 J of energy to raise the temperature of 1 kg of water by 1°C. Had Louganis fallen into only 1 kg of pool water, he would have had almost enough kinetic energy to raise the water temperature by a full degree. Fortunately, there were at least 2500 m^3 of pool water for Louganis to land in. With a mass of 2.5 million kilograms, the temperature of the water in an Olympic pool scarcely increases when Louganis enters it. The fact that it *does* increase means that Louganis's kinetic energy has dissipated to the pool. Energy is conserved.

But, you might say, why can't the pool transfer the lost 3750 J of kinetic energy back to Louganis? Why could he not recover enough energy from the pool to be thrown back up to the springboard? After all, those processes *do* conserve energy! You would be right about energy conservation. The first law of thermodynamics tells us that if we account for all internal energy changes, heat transfers, and work performed, we end up with conservation of energy. The first law of thermodynamics cannot, however, tell us anything about the *likelihood* that a given process will occur. Though it is true that if energy from the water were to revert to Louganis, propelling him out of the water and back onto the springboard, energy would be conserved, we all agree that if we saw such a thing, we would think we had lost our minds. We know that nature does not work that way.

The second law of thermodynamics solves our problem. It tells us that the total entropy of a closed system cannot decrease during a given process. In a loose and qualitative sense, entropy measures the amount of disorder in a system. During a given process, nature moves toward more disorder, but this concept is sometimes tough to visualize. You are never going to walk down a sidewalk and see six stones stacked on top of each other—stones that got that way because heat energy was taken up from the ground and given to the stones. The stacked stones represent order. You could certainly see six stones stacked up, but someone would have stacked them before you saw them. Although a person could give local order to the stones, the act of stacking them actually causes more disorder, since the person burns energy to perform the task. If Louganis were to pop out of the pool, order would increase, because the pool would be a tad cooler and there would not be as many waves in the water. The second law of thermodynamics says that it is impossible for such an event to take place, even though energy would be conserved.

We now leave the pool and look at a sport with a fan base numbering in the billions!

7

Soccer Kicks Gone Bananas

Off-Center Kicking and the Magnus Force

When the World Stops

Each January (or possibly early February), sports fans in the United States get geared up and excited for the Super Bowl. After the two conference title games provide the combatants for the Super Bowl, a fortnight of hype and analysis and silliness leads up to the big game. In the days following the Super Bowl, news outlets report television ratings that place the game among or near the top ten most-watched American shows in the history of television. There are also reports about how many people watched the game around the world, typically a couple of hundred million. The 2008 Super Bowl, in which the Giants defeated the Patriots in thrilling fashion, had a Nielsen-reported average audience of around 98 million in the United States alone.[1] For someone living in the States, like me, it's easy to get caught up in the hype and ratings and come away with the impression that the entire world stops what it's doing when the Super Bowl commands the screen. When I started graduate school, I had the good fortune to make friends with people from outside the States. They would see me get worked up for the Super Bowl and offer either puzzled looks or wry smiles. It dawned on me then that perhaps the world really doesn't stop when the champion of the National Football League

---

[1] Nielsen also reported that about 148 million in the United States—half of the nation's total population—saw at least a portion of the 2008 Super Bowl. That game turned out the be the second-most-watched show in U.S. television history, behind only the final *M*A*S*H* episode in 1983, which had an average U.S. audience of about 106 million people.

is crowned. The wry smiles I saw hinted at something else, too. I simply didn't get it back then.

The reality is that the number of people who watch and care about American football is dwarfed by the number of people in the world who watch and care about another type of "football," what those in the United States call *soccer*. When it's World Cup time, the world (sans us) really does stop! In parts of Europe, for example, shops close and businesses take a break when fellow countrymen are in action. Many in the States probably cannot imagine how passionately some in this world take soccer. The Super Bowl is big, but it does not rouse *billions* of people the way the World Cup does.

I turn now to the physics of soccer. After all, the wide world of sports physics would not be remotely complete without a visit to the soccer field. We have already discussed the flight of projectiles, be they footballs or people. When thinking about what makes a soccer ball's flight special, we will consider some new physics. The world may stop spinning when the World Cup games are underway, but the soccer ball's spin helps it elude even the most gifted goalkeepers.

Beckham Goes Bananas!

Before he came to the Los Angeles Galaxy in 2007 for a cool quarter-of-a-billion-dollar five-year contract, David Beckham had gained international celebrity as one of the best footballers England had ever produced. He was twice a runner-up for FIFA's[2] prestigious "World Player of the Year" Award. Brazil's Rivaldo beat Beckham for the award in 1999, and Luís Figo of Portugal took the award in 2001. Before you imagine that second place is not so hot, think about the fact that the award is given to the best soccer player *in the world*. No other Englishman has done better than Beckham's two second-place finishes.

Beckham is also the only Englishman to score a goal in three World Cup competitions. That's not bad, considering that the World Cup began

---

[2] FIFA stands for the Fédération Internationale de Football Association, or what those in the United States call the International Federation of Association Football. It is the organization that governs international soccer.

play in 1930 but is played only every four years. Beckham's third World Cup goal came in 2006 on a free kick against Portugal in a quarter-final match, which ended in a tie. He also scored a World Cup goal in 1998 (a free-kick goal) and a goal in 2002. Beckham's free-kick abilities became legendary after the release of the 2002 British film *Bend It Like Beckham*. Even though the movie was not focused on David Beckham, the title alone gave people around the world the idea that he could do something special with a soccer ball. There are certainly scores of others out there who can do magical things with a soccer ball—Brazil's Roberto Carlos, hardly déclassé to Beckham, has made some kicks that have made my head spin. But it's Beckham who got his name in the movie title.

So, what is it that Beckham "bends"? The soccer ball, of course, bends a little whenever someone kicks it. But that's not what *bending* a soccer kick means. We have already seen how balls and people soar through the air. With a constant gravitational force and no air resistance, the trajectories are parabolas. (I showed you parabolic paths in, for example, Figures 3.3 and 6.6.) Air resistance alters the path (it is no longer a parabola), but the shape is not much altered. But whether we include the air or not, the paths we have examined to this point have been *two-dimensional*. That means that if we turn the path 90° in, say, Figure 3.3 and look end-on, we would see only the projectile rising and falling. All the motion, that is, would be along a line. What Beckham and many other footballers are able to do is to impart *spin* on the soccer ball when they kick it. So long as the spin axis is not pointing exactly to the left or right as seen by the kicker, the ball can move in a *third* dimension. Looking end-on, as before, we would see the soccer ball move away from its straight-line motion as it moves either to our left or to our right (depending on the spin direction). The trajectory is thus "bent" away from what we would normally perceive. (Bent kicks in soccer are sometimes called "banana kicks" because the trajectory looks a little like a tilted banana; see Figure 7.1.)

There are numerous David Beckham highlights out there. I urge you to watch some of his free kicks on YouTube. Watch clips of other great soccer greats, too. Information and media are much more readily available in today's world than in years past. Whether you find a clip of Beckham scoring for Manchester United, Real Madrid, or England's national team in World Cup play, you will see some amazing trajectories. The free-kick highlights are especially fun to view. You see the soccer ball at rest on the

Figure 7.1. The trajectory of a bent soccer kick looks a little like a banana.

ground; a wall of defenders readies for the kick; an anxious goalie awaits the chance to thwart the great Beckham; the crowd is going nuts. Beckham kicks; the ball sails along its banana-like path; the wall of defenders flinches; the goalie, left out of position, is helpless against the perfect kick. Goal! The crowd goes bananas; Beckham slides along the ground; teammates mob their star. For a split second, the world ceases to spin!

Free Kicks and Corner Kicks

In the United States, the "big three" dominate the sports scene. Baseball, football, and basketball have always garnered the majority of American sports enthusiasm. We tend to gravitate toward what we are most familiar with—for more than a century, baseball has been known as the "national pastime"—and soccer has not captured the American sports fan like it has the sports fans in the rest of the world. Some in the States demean soccer because there is so little scoring. Without having played the game in their youth, most Americans simply do not understand the nuances and intricacies of soccer. I concede that I came to appreciate the phenomenal athleticism and skill of elite soccer players only in my mid-20s. When I watched the United States lose to Brazil on Independence Day (July 4th) in 1994, even my novice eyes told me that the eventual World Cup champion Brazilian team was far better than my country's squad. If you saw the score only in the newspaper, you might have thought the match close. The 1-0 score hid most of the action.

Consider a game like basketball. Say that a National Basketball Association game ends with a score of 100–92. There are 48 minutes in a regulation NBA game, meaning that in our hypothetical game the average rate of scoring was four points per minute. That works out to about a basket per team for each minute of action. With the 24-second shot clock (instituted in 1954) and nearly 200 points in most games, we see a lot of scoring in a basketball game, compared to what we see in a soccer match. If a basketball team gets down by six or eight points early in the game, coaches change little, if any, of their game strategy. In soccer, however, an early goal or two can change drastically how a team plays. Lacking a shot clock, teams in the lead may try to stall while in possession of the ball. Because scoring opportunities are so much rarer in soccer than in a game like basketball or football, tension rises precipitously when teams are awarded a free kick or a corner kick.

When a player commits a foul or touches the ball with one or both hands, the opposing team is awarded a free kick, those taken within about 32 m (~35 yards)[3] from the opposing goal, having the best chance for a goal. The kicker places the ball about 9.14 m (~10 yards) from the players on the other team. A few defenders typically constitute a "wall" wherein they basically prepare themselves to be struck by the ball if they are "lucky" enough to get in the way of the kick. The goalie must guard a rectangular-shaped goal that is 7.32 m (8 yards) wide and 2.44 m (8 feet) tall. That nearly 18 m^2 (192 ft^2) of area is a bit bigger than you might think when you watch soccer on television. See Figure 7.2 for a sample free-kick player configuration. Because a kicked ball is typically in the air for about a second, the goalie is hard-pressed to cover 7.32 m (8 yards) in that time, especially since his/her reaction time will account for as much

[3] My decision for this chapter's units system was a difficult one. Much of the distances with which I am familiar are stated in imperial units like yards and feet. Most of the world, and essentially all of the world's scientific community, uses the SI system of units. I chose here to go along with the majority, though I will provide conversions. Since my natural way of thinking about soccer is in imperial units, the first numbers I came up with are in that system. That is why some of my SI units may at first seem a little odd.

Figure 7.2. The Lynchburg College men's soccer team practices a free kick. Note the wall of defenders between the ball and the goal. (By permission of Lynchburg College)

as half a second. To compensate for that reaction time, goalies will often guess the kicker's intended target by moving in a chosen direction just as the opponent kicks the ball.

If a defender is the last to make contact with a ball that goes out of bounds by crossing the goal line, the opposing team is then awarded a corner kick. A player kicks the ball from one of two quarter-circles of radius 0.914 m (1 yard) that are located at the corners of the field on both sides of the goal. See Figure 7.3 for a sample corner-kick player configuration, the kicker in the foreground. We have seen that the kicker with a free kick is aiming directly for the goal. A player kicking a corner kick is not usually aiming for the goal. Since the plane of the goal is not in front of the kicker, but almost 90 degrees removed, the player often aims for a region in front of the goal where a teammate could kick, head, or (for more flair and excitement) bicycle kick the ball into the goal.[4]

[4] A player executes a bicycle kick by jumping up while rotating backwards—almost like a somersault—and kicking the ball at a point above the ground roughly comparable to a player's height. The legs appear to be pedaling during such a kick.

Figure 7.3. The Lynchburg College men's soccer team practices a corner kick. The right-footed kicker aims for a region in front of the goal. (By permission of Lynchburg College)

I want now to describe how a soccer ball's spin can make it follow an arc suggesting a banana lying on its side. I also want to present a model of free kicks and corner kicks that allows one to predict success rate. Let's move on to the physics!

Physics Time

Think about all of the various types of balls used in sports: baseballs, footballs, soccer balls, basketballs, cricket balls, rugby balls, golf balls, and so on. Each ball has its own unique character and style. What most balls used in sports do have in common is some type of surface imperfection, meaning that none of the balls I've listed are perfectly smooth. Some balls, like baseballs and footballs, have prominent stitches that hold the balls together. It's doubtful, of course, that the first person to stitch up a baseball had fluid mechanics in his or her mind. But were it not for those wonderful 108 double stitches on a baseball, home runs in today's parks would be almost nonexistent. Many years ago, the aristocracy used smooth balls in the game of golf, and routinely discarded used balls. Less affluent golfers would settle for the used balls and note that the nicked and cut-up balls went farther than the new, smooth ones. Today, golf balls are made with enough dimples to make them as "rough" as possible. Without those dimples, Tiger Woods and his fellow golfers could not whack those 300 + yard (274 + meter) drives; their drives could not make it half as far.

Though it may at first seem contradictory, balls with surface roughness often experience *less* air drag than smooth balls, at least at the speeds associated with most sports. Imagine a nonspinning ball moving through the air. The ball has to "push" air out of the way to get through it. We already discussed, back in chapter 4, how Lance Armstrong had to push his way through the French air. When discussing air resistance with respect to Armstrong, I didn't go into much detail because of the enormous complexity of air's interaction with a person on a moving bike. And even an object as seemingly simple as a smooth sphere requires scientists to employ sophisticated computer models in describing the phenomena of air resistance. The addition of stitches or other surface imperfections further complicates the matter.

Complicated computer models are beyond the scope of this book; even most scientists don't get too worked up over reading about them. Let's stay qualitative for now, and later I'll use a fairly simple model we have already discussed. If you push on me, I must push on you, as prescribed by Newton's third law. Well, if a ball has to move air out of the way, the air, for its part, must feel a force from the ball, meaning that the ball feels a force from the air. Because of surface friction and other complications, a *boundary layer* of air forms on the ball as the ball proceeds on its trajectory. That thin layer of still air around the ball keeps incoming air from reaching the ball's surface. It's almost as if the ball has a tiny air blanket around much of its surface. (If you've ever wondered why the accumulated dust on a fan does not get "blown off" when the fan is on, blame the boundary layer.)

Understanding the properties of the boundary layer is crucial to determining the air's influence on the ball. It was not until the turn of the twentieth century with the work of the German physicist Ludwig Prandtl that people began to understand the importance of the boundary layer to

Figure 7.4. A sketch of a soccer ball moving through the air, but viewed in the ball's reference frame. The ball moves from left to right, meaning that the air moves from right to left from the ball's point of view. Note the many eddies behind the ball. The air flow is "laminar."

fluid mechanics.[5] When the boundary layer sheds from the back portion of the ball, a wake of eddies is created. See Figure 7.4 for a sketch of this phenomenon. All that kinetic energy in the eddies is evidence that the ball has lost energy to the air, and thus we have a simple and relatively crude idea of how air resistance works.

I need to incorporate air resistance into my model of a soccer ball's flight, but I don't want the model to be so sophisticated that you decide to free kick this book. Let's use the model for air drag we have already seen, namely equation (4.2), now remembered as

$$F_D = \tfrac{1}{2}\, C_D\, \rho\, A\, v^2. \tag{7.1}$$

What really makes using this equation difficult is that C_D depends on the soccer ball's speed. Above about 8 m/s ($\sim$18 mph), the drag coefficient drops to about 40% of its low-speed value. Figure 7.5 shows experimental data for the drag coefficient as a function of flight speed. The data are taken from Matt Carré's research group at the University of

---

[5] A nice, easy-to-follow account of the history and the physics of Prandtl's work may be found in "Ludwig Prandtl's Boundary Layer," by John D. Anderson, Jr., in the December 2005 issue of *Physics Today* (pp. 42–48).

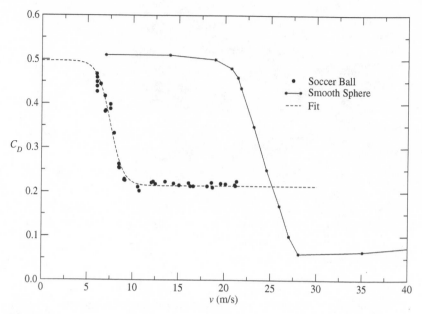

Figure 7.5. Drag coefficient C_D versus flight speed v for a model soccer ball. The dashed line is equation (7.2). Note that $10\,\text{m/s} \simeq 22.4$ mph. (Soccer data courtesy of Matt Carré. The smooth sphere data come from Achenbach's work.)

Sheffield in the United Kingdom.[6] For comparison, Figure 7.5 also shows drag coefficient data for a smooth sphere.[7] Note that the rougher soccer ball has a smaller drag coefficient than a smooth sphere. Above the so-called "critical speed," after the air flow goes from "laminar" to "turbulent," turbulence in the boundary layer actually delays the boundary layer's separation from the ball. (See the sketch in Figure 7.6 and compare with the sketch in Figure 7.4.) The drag force itself, given by the product of everything on the righthand side of equation (7.1), continues to increase with speed until the ball begins to near the critical speed. Figure 7.7 shows experimental data for the drag force on a soccer ball and comparison data with a smooth sphere. In the vicinity of the critical speed, the drag force F_D

[6] For more details about the soccer data in Figures 7.5 and 7.7, see "Understanding the effect of seams on the aerodynamics of an association football" by Carré, Goodwill, and Haake in *Mechanical Engineering Science*, volume 219, pp. 657–66 (2005).

[7] The smooth sphere data are taken from the classic article "Experiments on the flow past spheres at very high Reynolds numbers" by E. Achenbach in the *Journal of Fluid Mechanics*, volume 54, pp. 565–75 (1972).

Figure 7.6. The soccer ball in the sketch here moves faster than the one in Figure 7.4. Note the delay in the boundary layer separation. The air flow is turbulent.

increases more slowly than v^2, where v is the speed of the ball. Only for ball speeds not close to the critical speed can we claim roughly that $F_D \propto v^2$.

The experimental data may be fitted to the following function:

$$C_D = a + \frac{b}{1 + \exp\left[(v - v_c)/v_s\right]}, \qquad (7.2)$$

where $a = 0.214$, $b = 0.283$, $v_c = 7.50\,\text{m/s}$, and $v_s = 0.707\,\text{m/s}$. The form of equation (7.2) is motivated by an equation that Nicholas Giordano developed for a baseball's drag coefficient.[8] I put equation (7.2) on Figure 7.5. For spheres, C_D approaches 0.5 as the speed goes to zero. My fitted function goes to about 0.497 at zero speed. The a coefficient represents the large v limit. The speed v_c is roughly where the critical speed is located. The parameter v_s is the scaling speed in the exponential function.

For more sophisticated work, I would use equation (7.2), so that I would have a value of C_D at all speeds that is reasonably consistent with experimental data. To simplify our work here, however, I will take $C_D = 0.275$, i.e. a constant. That estimate, based on available

[8] See *Computational Physics* (2nd edition) by Nicholas J. Giordano and Hisao Nakanishi (Pearson/Prentice Hall, 2006), p. 33.

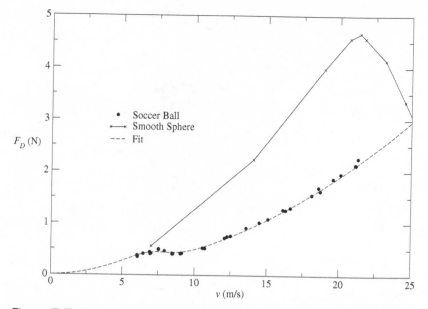

Figure 7.7. Drag force F_D versus flight speed v for a model soccer ball. The dashed line comes from plugging equation (7.2) into equation (7.1). Note that $1\,\mathrm{N} \simeq 0.225$ pounds. (Soccer data courtesy of Matt Carré. The smooth sphere data come from Achenbach's work.)

experimental data, is a reasonable average.[9] I still need a computer to determine the trajectory, even with a constant value for C_D.

I once again take the air density to be $\rho = 1.2\,\mathrm{kg/m^3}$. The rules state that a soccer ball must have a circumference of 68 cm to 70 cm (27 in to 28 in). For a ball with a circumference of 27.5 in (nearly 70 cm), the radius is about 0.111 m, meaning its cross-sectional area is $A \simeq 0.0388\,\mathrm{m^2}$ ($\sim$60.2 in^2). We now have all we need to use equation (7.1) for the air drag on a soccer ball.

Imagine now that David Beckham approaches the ball and then kicks it with his foot slightly off-center. Because the line of action of the force from his foot does not pass through the ball's center of mass, the ball experiences a torque. Beckham's foot may also slide briefly along the

---

[9] In addition to the Sheffield group's data, Bray and Kerwin's "Modelling the flight of a soccer ball in a direct free kick" in a 2003 issue of the *Journal of Sports Science* (volume 21, pages 75–85) helped me with my estimate for C_D.

Figure 7.8. A sketch of a soccer ball spinning clockwise. The ball moves from left to right, meaning that the air flow is right to left. Note how the boundary layer separation is delayed on the bottom of the ball.

ball's surface, allowing friction between his shoe and the ball to generate more torque. The torque creates angular acceleration as the ball goes from rest to some maximum angular speed, once Beckham's shoe loses contact with the ball. We must now consider what happens when the ball is spinning.

But first, remember that friction between the air and the ball is responsible for the boundary layer. Note in Figure 7.4 that the points on the sides of the soccer ball, where the boundary layer has not separated, remain symmetric. If instead the ball is spinning, it whips air around itself and delays one of the separation points. Figure 7.8 shows a clockwise-spinning ball moving left to right, into the ambient. The air at the bottom of the ball gets whipped underneath, thus delaying the boundary layer separation on the bottom of the ball. The air at the top of the ball is met with the ball spinning right into the air, meaning that the separation point occurs earlier there than what Figure 7.4 shows. Since the ball deflects the air upward, Newton's third law tell us that the air must deflect the ball downward. (The same idea explains how a boat rudder is able to help make a boat turn in the water. If the rudder deflects water in one direction, the rudder must feel a force in the opposite direction.)

We now have a new force to consider, in addition to the ball's weight and the drag force. Like the drag force, this new force clearly would not exist if all of this were to be happening in a vacuum. I can motivate a form for an equation for this new spin force in the same way I motivated equation (7.1) for air drag. Air density, cross-sectional area of the ball, and ball speed certainly influence the spin force. It should also be easy to see that the ball's rotational speed should play a role, too.(For Beckham, a critical role.) Analogous to equation (7.1), I write the new spin force as

$$F_M = \tfrac{1}{2} C_M \rho A r \omega v, \qquad (7.3)$$

where r is the soccer ball's radius, ω is the rotational speed (in rad/s), and C_M is a dimensionless coefficient analogous to C_D.[10] For simplicity, I will do as I did with C_D and take C_M to be constant, even though C_M may also depend on v. In my model, $C_M = 1$. The "M" label is due to the name of the person for whom the spin force is named. We call this new force the "Magnus force" after the German scientist Heinrich Gustav Magnus (1802–1870). Magnus published "On the deviation of projectiles; and on a remarkable phenomenon of rotating bodies" in Berlin in 1852; an English translation appeared in London the following year. Even though Magnus has had the honor of having his name on the effect, it was Newton who, long before, actually noticed the curved path of spinning tennis balls.[11] Whether Magnus knew that of Newton or not I don't know. We'll go ahead and use the term "Magnus effect." Newton has his name on enough stuff already!

Obtaining the direction of the Magnus force requires the right-hand rule. Recall from chapter 5 that finding $\vec{\omega}$ requires curling one's right-hand fingers around with the rotating portions of an object, the thumb pointing along $\vec{\omega}$. In Figure 7.8, the right-hand rule tells us that $\vec{\omega}$ points into the page. After finding $\vec{\omega}$, we need to use the right-hand rule one more time to find $\vec{F}_M$. I promise that we need only use the right-hand rule twice! Point the fingers of your right hand along $\vec{\omega}$; then, swing them toward the ball's velocity, $\vec{v}$. In Figure 7.8, the vector $\vec{v}$ points to the right. Your right

---

[10] Equation (7.3) is sometimes written as $F_M = \tfrac{1}{2} C_L \rho A v^2$, where C_L is the "lift coefficient." If speeds are not too small, and if angular speeds are not too large, C_L is roughly proportional to $r\omega/v$; the proportionality factor is C_M.

[11] See the fine book *Isaac Newton* by James Gleick (Pantheon Books, 2003), pp. 79 and 90.

hand may be a little twisted, but your right thumb should be pointing straight down. That is the direction of $\vec{F}_M$.[12]

We now have all the ingredients we need to model the flight of a soccer ball. The net external force on the ball while in flight consists of the ball's weight, the air drag, and the Magnus force. The vector sum of all those forces equals the soccer ball's mass times its acceleration vector—Newton's second law. If the ball has sidespin (has been "bent"), the motion takes place in three dimensions.

Time for a quick baseball aside! If you have ever stood in a batter's box and tried to hit a slider, you know that three-dimensional motion can be tricky. A baseball pitcher throws a slider with a great deal of sidespin on the ball. A right-handed batter looking at a right-handed pitcher sees a slider move left to right in addition to a component of vertical motion created by gravity and air drag. Note, too, that the Magnus force applies for any orientation of a spinning ball. A baseball pitcher throws an over-hand curve by pulling two fingers down on the front of the ball just before release. The batter sees the front of the ball moving down, meaning the Magnus force points down. This is the "fall off the table" curveball. A fastball spins in the opposite direction, meaning the Magnus force is up. The "rising fastball" does not really rise; it merely drops less than it would if gravity and air drag were the only forces acting on the ball.

Footballers like David Beckham impart sidespin on a soccer ball to impart a side-to-side motion that often eludes defenders and goalies. The sideways curve leads to those wonderful banana-like trajectories that cause our heads to spin and make us ask, "How does Beckham bend it like that?!"

Some Warm-Up Numbers

Before getting to the problem of interest, I wish to give you a feeling for some of the numbers involved. We established the three forces on the

---

[12] More mathematically speaking, $\vec{F}_M$ points in the direction of $\vec{\omega} \times \vec{v}$, where "$\times$" is the "vector product" or "cross product." I will not be discussing the "reverse Magnus effect" in this book, for which special circumstances allow for $\vec{F}_M$ to point toward $-\vec{\omega} \times \vec{v}$.

soccer ball in the preceding discussion. The ball's weight is

$$W = mg \simeq (0.425\,\text{kg}) \left(9.8\,\frac{\text{m}}{\text{s}^2}\right) \simeq 4.2\,\text{N} \simeq 0.94\,\text{lbs}. \qquad (7.4)$$

Let's suppose that Beckham kicks the ball with an initial speed of 25 m/s (~56 mph)—a strong kick. Equation (7.1) gives the drag force as

$$F_D \simeq \tfrac{1}{2}(0.275)\left(1.2\,\frac{\text{kg}}{\text{m}^3}\right)(0.0388\,\text{m}^2)\left(25\,\frac{\text{m}}{\text{s}}\right)^2 \simeq 4.0\,\text{N} \simeq 0.90\,\text{lbs}. \quad (7.5)$$

Note that when first kicked, the drag force on a soccer ball is comparable to its weight. Suppose now that Beckham imparts a spin on the ball of 10 revolutions per second (~62.8 rad/s). Equation (7.3) gives the Magnus force as

$$F_M \simeq \tfrac{1}{2}(1.0)\left(1.2\,\frac{\text{kg}}{\text{m}^3}\right)(0.0388\,\text{m}^2)(0.111\,\text{m})\left(62.8\,\frac{\text{rad}}{\text{s}}\right)\left(25\,\frac{\text{m}}{\text{s}}\right)$$

$$\simeq 4.1\,\text{N} \simeq 0.91\,\text{lbs}. \qquad (7.6)$$

The Magnus force, too, is comparable to the soccer ball's weight. Air drag and the Magnus force thus play significant roles in determining the trajectory of a soccer ball. Note that only the ball's weight remains fixed throughout its trajectory. Air drag slows the ball down, meaning that both F_D and F_M decrease.[13] Air friction will also create a torque on the ball that reduces the ball's angular speed. I will ignore that complicated effect and suggest that the ball's rotational speed does not change terribly much during its flight. I will thus keep ω fixed in my calculations.

What Are the Odds? — Free Kicks

What do you think the odds are for a player like David Beckham to score a goal on a free kick? What about the odds for a player kicking a corner kick to the optimum spot for a teammate to score a goal? I want to use the results of some experiments and some good ol' physics to answer those questions. I'll tackle the first question here and then look at corner kicks.

---

[13] Accounting for the speed dependence of C_D means that there is a range of speeds for which there is a slight *increase* in F_D as the speed *decreases*. See Figure 7.7 around 8 m/s–9 m/s. Although that behavior interests me greatly, I will keep things simpler here and stick with a constant C_D.

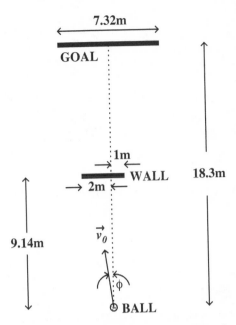

Figure 7.9. Overhead view of free-kick geometry. The wall is as wide as the ball, and its height is 1.83 m (6 ft.). The goal's height is 2.44 m (8 ft.). Note that the launch angle, θ, is not shown, and that the approach angle, ϕ, is positive for counterclockwise measurements.

As I had to do earlier in this book, I need to use a computer to solve Newton's second law for the trajectories. But I won't bore you with the details of the computer code.

There are numerous places on the field where a footballer could kick a free kick. I will simply pick one spot and go from there. Figure 7.9 shows an overhead view of my model's free-kick geometry. David Beckham will kick the ball 18.3 m (20 yards) from the center of the goal. A 3 m wide wall of defenders is 9.14 m (10 yards) in front of Beckham, halfway to the goal. The wall is set off left of center, in an attempt to guard the left portion of the goal. Figure 7.10 shows where the goalie stands. The idea is that the wall tries to protect the left portion of the goal while the goalie tries to protect the right portion. The "target" in the upper-left portion of the goal is where Beckham wants the ball to go. The goalie will most likely block a ball headed toward the lower-left portion of the goal because the ball will be in the air long enough (greater than about 1 second) for the

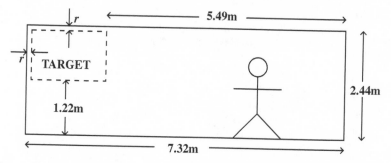

Figure 7.10. Front view of the goal for the free-kick model. The goalie is 1.83 m (6 ft.) tall. The goal is 7.32 m (8 yards) wide and 2.44 m (8 ft.) tall. The target is in the upper-left corner of the goal, offset from the goal pole by the ball's radius r.

goalie to overcome his reaction time and get to the left edge of the goal in time. Beckham must clear the wall of defenders and sneak the ball into the target to score a goal.

Beckham will essentially have four parameters at his disposal. He can control the launch speed, v_0, by how hard he kicks the ball. He puts a certain angular speed, ω, on the ball, its value determined by how far left or right of center his foot strikes the ball. How far below the ball's center of mass he strikes the ball will determine the launch angle, θ, measured from the ground (not shown in Figure 7.9). Finally, Beckham approaches the ball at some angle ϕ, measured from a line drawn from the ball to the center of the goal.

What I want to do now is consider ranges of Beckham's four parameters. No soccer player, not even David Beckham, is capable of setting each of the four parameters exactly how he or she wishes. There will always be some range of possibilities in where and how a soccer player strikes a ball. Drawing upon experimental studies,[14] I chose the following ranges for Beckham's free-kick parameters: 24.6 m/s (55 mph) $\leq v_0 \leq$ 29.1 m/s (65 mph); 6 rev/s $\leq \omega \leq$ 12 rev/s, 13° $\leq \theta \leq$ 17°; and 1° $\leq \phi \leq$ 5°. Since Beckham is right-footed, I take the soccer ball to spin

[14] I used the work by Bray and Kerwin mentioned in footnote 9 and two other papers: "The curve kick of a football I: impact with the foot" by Asai, Carré, Akatsuka, and Haake in *Sports Engineering*, volume 5, pp. 183–92 (2002), and "The curve kick of a football II: flight through the air" by Carré, Asai, Akatsuka, and Haake in *Sports Engineering*, volume 5, pp. 193–200 (2003).

counterclockwise as seen from above. Though it is easy to change the orientation of $\vec{\omega}$ in my code, I will simply assume that $\vec{\omega}$ points up in all my calculations.

Instead of thinking about a single trajectory, I want to think about *millions* of them. Since computers are superb for calculating fast and doing many calculations over and over again, I want to calculate many trajectories for the aforementioned ranges, and then see which ones hit the target in Figure 7.10. I need step sizes, or increments, for my ranges, so that my computer knows which trajectories to calculate. I chose the following: $\Delta v_0 = 0.0447$ m/s (0.1 mph), $\Delta \omega = 0.1$ rev/s, $\Delta \theta = 0.1°$, and $\Delta \phi = 0.1°$. I assume that Beckham is equally likely to strike the ball for all values of the four parameters in the given ranges. (This is the so-called "equal a priori probabilities" assumption.) If you count how many steps, or increments, I have for each of the four parameters, you will get 101 v_0 values, 61 ω values, 41 θ values, and 41 ϕ values. (Don't forget to count the beginning and ending values!) That means that my computer will calculate a total of $101 \times 61 \times 41 \times 41 = 10{,}356{,}641$ different trajectories. It's nothing for a computer to solve Newton's second law ten million times!

After all trajectories have been determined, I simply ask the computer to inform me how many of them struck the desired target. The ratio of (kicks that hit the target) to (total kicks) will then give me an estimate of the odds of a player like David Beckham successfully kicking a free kick. I found that about 9.63% of all kicks hit the target. In other words, the odds for a successful free kick, even for Beckham, are about 1 in 10. That figure agrees at least qualitatively with what I have observed when watching soccer games. There are not, however, exhaustive statistics for soccer, like those for, say, baseball.[15]

To illustrate what I have just done, consider each kick as having its own unique set of parameters $\{v_0, \omega, \theta, \phi\}$. Each kick then represents a point in a four-dimensional space. "Ugh!" you might ask, "what the heck are you talking about with *four*-dimensional spaces?!?" Math and science geeks like me can talk about any number of dimensions we like.

[15] There is no comparison here to the statistics kept in baseball! If you want to know how a given pitcher does against left-handed hitters on the road in day games, you can find out without too much trouble. As an amateur sabermetrician, I love baseball statistics, and I wish I could find good statistics for soccer. If you know where I can get free kick stats, please drop me an e-mail.

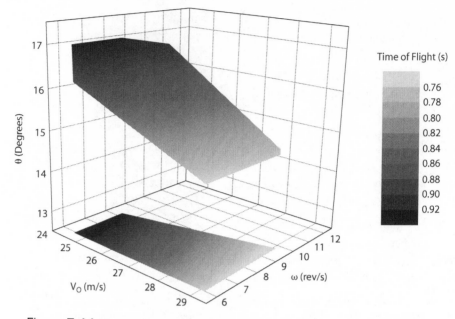

Figure 7.11. Parameter space of successful free kicks for $\phi = 3°$. The flight times are shown on the right. The shading at the bottom is given merely to help one visualize the shaded volume.

I certainly cannot *visualize* four dimensions in my head, but I can talk about them and calculate with them. Instead of trying to do anything fancy with four dimensions, let's stick with something more familiar, like three dimensions. Check out Figure 7.11. The three-dimensional rectangle defined by the coordinate axes represents the "volume" of all kicks for $\phi = 3°$. I picked ϕ to be constant because I thought that that parameter is the easiest for Beckham to control. The shaded object in the figure represents the "volume" of all kicks that hit the target. Note, though, that that is a three-dimensional "slice" of the entire four-dimensional space. I would need to draw Figure 7.11 for each of the 41 values of ϕ in its range. What Figure 7.11 does show is the amount of volume occupied by the "good kick" shape for the specific choice of $\phi = 3°$. Note, too, in the figure that times of flight are all under one second. Anything more than that and a goalie will have a good chance to get to the ball.

One of the successful kicks in Figure 7.11 is shown in Figure 7.12. Note how tricky it is to put sufficient spin on the ball! By the time the goalie sees

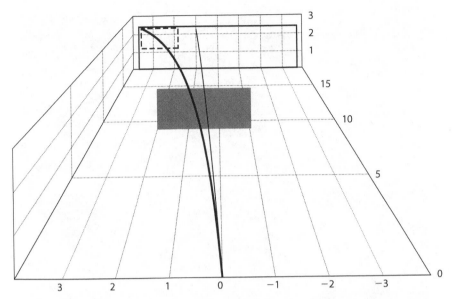

Figure 7.12. A successful free kick from Figure 7.11. The heavy line shows a free kick for $v_0 = 28.6$ m/s (64 mph), $\omega = 9$ rev/s, $\theta = 15°$, and $\phi = 3°$. The light line is what the same kick would look like with no spin ($\omega = 0$). All distances are in meters.

the ball coming over the wall of his teammates, it is curving away from him. Had the ball not been spinning, the goalie probably would block the ball, since it would head for the goal just left of center. Beckham's banana kick bends right into the upper left corner of the goal for a score! (Occasionally, at least.)

What Are the Odds? — Corner Kicks

When thinking about the odds for a good corner kick, we need to think about the ball's reaching a teammate who is in a good position for a goal. Though it *is* possible to score on a corner kick, it is highly unlikely (See Figure 7.13.). The strategy is to set one's teammate up for the glory. Figure 7.14 shows an overhead view of the corner kick geometry. My model's "target" in this case is an imaginary rectangular box (parallelepiped) sitting 1.68 m (5.5 ft.) above the ground. Its dimensions are 7.32 m

Figure 7.13. The next Mia Hamm? Clare Lipscombe achieved the rare distinction of scoring directly into the goal from a corner kick. (Courtesy Trevor Lipscombe)

(8 yards) × 1.83 m (2 yards) × 0.762 m (2.5 ft.). The top of the box is thus 8 ft. above the ground. The idea is that the kicker aims for an area where a teammate can head, kick, or bicycle kick the ball into the goal. I chose the target to be as wide as the goal, but offset by 0.914 m (1 yard), as shown in the figure. The kicker will send the ball into the target region, the ball moving up and to the left, as the direction implied by the short arrow at the lower right would indicate.

There are, of course, defenders in the region as well. The success of the footballer in my model is determined solely by his or her ability to kick the ball into the target. The odds of a goal then shift to the teammate who kicks or heads the ball. The odds I compute here are for the kicker getting the ball into the target, not the poorer odds on whether or not a goal is scored subsequently by a teammate.

I now do as I did with free kicks; I choose a range for the four parameters the footballer can control. For the corner kick, I chose: 24.6 m/s

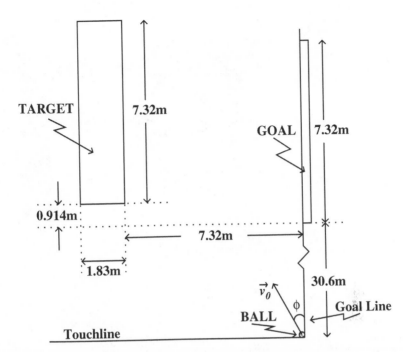

Figure 7.14. An overhead view of the corner kick geometry (not to scale). The target is a rectangular box (parallelepiped) sitting 1.68 m (5.5 ft) above the ground, with dimensions 7.32 m (8 yards) × 1.83 m (2 yards) × 0.762 m (2.5 ft.).

(55 mph) $\leq v_0 \leq$ 29.1 m/s (65 mph); 3 rev/s $\leq \omega \leq$ 9 rev/s; 24° $\leq \theta \leq$ 28°; and $-2° \leq \phi \leq 2°$. I figured the speed range would be about the same, so I left the range for v_0 alone. Because the ball has to travel a greater distance, I lowered the angular speed values, assuming the footballer would want a little more control of the kick. The greater distance also means that I had to increase the launch angles. After all, the ball has to hit a target more than 32 m (35 yards) away and almost the height of a footballer above the ground. For ϕ, I have seen corner kicks with lots of spin that appeared to have headed initially behind the goal line. Harder kicks with less spin are aimed away from the goal line.

I chose the same step sizes, or increments, that I chose for the free kicks. Because all four parameter ranges encompass the same width as before, we have 10,356,641 total corner kicks. After running my code for the trajectories, I ask the computer to count how many of those kicks passed through the target. Roughly 24.5% of the total number of kicks

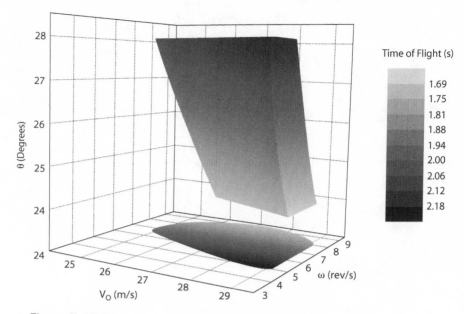

Figure 7.15. Parameter space of successful corner kicks for $\phi = -1°$. The flight times are shown on the right. The shading at the bottom is given merely to help one visualize the shaded volume.

were successful to that extent. That means that the odds of a good corner kick are about 1 in 4. Given a larger target, and a three-dimensional one at that, it should not be surprising to find that it is easier to hit a good corner kick into its target area than it is to score a goal directly, on a free kick. Bear in mind, though, that a "successful" corner kick may not result in a goal.

Again, I will give you a qualitative feel for which parameters work. Fixing the approach angle once again, this time at $\phi = -1°$, I can show you a three-dimensional slice of the four-dimensional parameter space. Figure 7.15 shows the volume of good kicks for my chosen value of ϕ. Figure 7.16 shows one of the successful kicks from Figure 7.15. Note that had there been no spin on the ball, it could have landed *behind* the plane of the goal, as shown in the figure. Spin allows the footballer to steer the kick more directly into the target so that it may be more successfully knocked in for a goal.

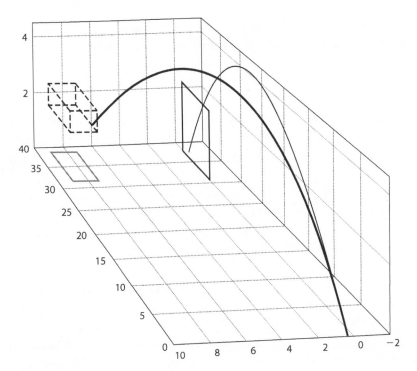

Figure 7.16. A successful corner kick from Figure 7.15. The heavy line shows a corner kick for $v_0 = 27.3 \text{ m/s} (61 \text{ mph})$, $\omega = 7 \text{ rev/s}$, $\theta = 24°$, and $\phi = -1°$. The light line is what the same kick would like with no spin ($\omega = 0$). All distances are in meters.

If you think statistics on free kicks are hard to find, forget trying to find any stats on the types of corner kicks I have described here. The 1 in 4 rule of thumb is, like the 1 in 10 rule of thumb for the free kick, reasonably consistent with what I have seen when observing soccer matches. If anyone wishes to watch a few hundred soccer games and compile some stats, you have at least one person interested in your findings!

Extended Time

One thing about soccer that bothers some Americans is that the players do not know exactly how much time is left in the game. There is no official game clock for fans to view. Referees may tack on extended time to account for player injuries, players goofing off, player substitutions, or anything else that referees believe have delayed the game.

There are so many more fascinating aspects of the game of soccer, and I simply do not have enough extended time, or space in this book, to touch on even half of them. I do hope, however, that using a little physics to determine some rules of thumb about free kicking and corner kicking gives you a glimpse of what physicists can do with the world's most popular sport. Even what I have done here is far from complete. There are many places on the field where free kicks may be taken, and there are many other strategies one can employ.

For example, one of Roberto Carlos's free kicks has never left my mind. In a 1997 match against France, the left-footed Brazilian placed the ball a little to the right of where I placed the ball in my model. He then swiped the ball below its center in a left-to-right manner, giving the banana trajectory a right-to-left motion, owing to the counterclockwise spin he had imparted to it (as seen from above). In other words, Carlos hit what those of us who have played baseball would call a "screwball." The ball passed to the right of the wall of defenders and sneaked into the far right side of the goal. The French goalie was not that far from the ball, but he looked like the deer caught in the headlights when that ball came around the wall. There are thus *many* modifications that could be made to what I have introduced to you here.

Brandon Cook, a former student of mine and an amateur soccer player, played a significant role in developing the model I have used in this chapter. If you are interested in reading more of the technical details, please see our paper.[16] It looks as if the referee is about to call an end to my extended time. Let's now go back to the Olympics. In fact, let's go back to *four* Olympics!

[16] "Parameter space for successful soccer kicks," by Brandon G. Cook and John Eric Goff in the *European Journal of Physics*, volume 27, pp. 865–74 (2006).

8

Four Olympics and *Four* Straight Gold Medals

Centripetal Motion and Lift

The REAL *Discobolus*!

We now make a final stop at the Olympics. Actually, we need to travel back to *four* Olympic games. Some may argue that what I have saved for here was the most remarkable of all Olympic performances. What images pop into your head when you think of *Olympics*? Do you think of ancient Greece, where the Olympics began nearly three millennia ago, in 776 BC? Does your mind conjure up an image of the five interlocking rings, first used in the 1920 Summer Olympics in Antwerp, Belgium? When I think of the Olympics, it is always Myron's *Discobolus* that comes to mind (see Figure 8.1). I'll be the first to admit that I am no art connoisseur, and I had to consult an art reference to learn who sculpted *Discobolus*. I learned that Myron did his most famous work in the middle of the fifth century BC. I did not need a lesson in Greek, however, to know that *Discobolus* means "discus thrower." One should be able to guess that at first glimpse of Myron's magnificent work! I can only dream about what was in his head when he worked on *Discobolus*. Was he thinking about one of the Greek Gods? Perhaps less poetic, he may have had in mind one of the local Athenian athletes of his time. But suppose we leave the plane of physical reality for just a moment and imagine that Myron possessed some Nostradamus-like power of clairvoyance.[1] If he could see a few thousand

---

[1] Michel de Nostredame, who those in the English-speaking world call Nostradamus, was a famous sixteenth-century "seer."

Figure 8.1. Myron's *Discobolus*, fifth century BC. (By permission of the Trustees of the British Museum)

years into the future, Myron surely had Al Oerter in mind when forming *Discobolus*.

Al Oerter was born on September 19, 1936, about a month after the summer games in Berlin ended. Most people know of Jesse Owens and his wondrous achievements that year. Owens's four gold medals there didn't sit too well with Adolf Hitler. Even the story of the German Luz Long giving helpful advice to Owens during the competition for the broad jump (now called the long jump) is known by many.[2] The historical significance of the Olympic setting in Nazi Germany, as well as the impressive feat of winning four gold medals in one Olympics, give legitimacy to Jesse Owens's status as an Olympic legend. Though Oerter was not confronted with the racial enmities Owens suffered, I believe his achievements rank with those of Owens and a handful of other Olympians in the pantheon of the "best of the best."

Like Owens, Oerter won four gold medals. Unlike Owens, who won gold in the 100 m, broad jump, 200 m, and 4 × 100 m relay, Oerter specialized in a single event—the discus throw. Since there is only one gold medal given out in each Summer Olympics for the discus throw, Oerter had to win the gold in four *different* Olympics. In fact, he won gold in four *consecutive* Olympics. Ponder that for a moment. Imagine training your butt off for a few years and becoming the best on the planet. Then do it again four years later, and then four years after that, and then again four years after that.

Even though Oerter won his last gold two years before I was born, meaning that I never had the privilege of watching him compete, I am continually astonished when I reflect on his accomplishment. When he passed away on October 1, 2007, at the age of 71, I saw a few stories about his life and achievements. Most people I know, however, are not even familiar with his name. If you forget much of the upcoming physics in this chapter, at least appreciate Al Oerter's achievement. Myron's *Discobolus* does not depict accurately how one actually throws a discus, but it does embody the grace and power of the Olympic ideal, and that ideal is Al Oerter (see Figure 8.2.)

Four in a Row

Opening ceremonies for the 1956 Summer Olympics in Melbourne, Australia, took place on Thursday, November 22. It was Thanksgiving Day

---

[2] Long was the silver medalist in the broad jump.

Figure 8.2. The great Al Oerter about to send the discus in flight at the 1956 Summer Olympics in Melbourne, Australia. (© Bettmann/CORBIS)

in the United States. While those in the northern hemisphere were getting ready for winter, those in the southern hemisphere were heading into summer. (We must thank the Earth, with its slightly more than 23° axial tilt, for our seasons!) Twenty-year-old Al Oerter had spent the previous two years throwing discus for the University of Kansas. His Olympic teammate, Fortune Gordien, was the world record holder in the discus at the start of the Melbourne games. Gordien's mark of 59.28 m (194 ft. 6 in.) had stood for more than three years. The American Sim Iness held the current Olympic record of 55.03 m (180 ft. 6 in.), set in his gold-medal-winning effort in Helsinki four years earlier. Oerter's gold-medal-winning throw of 56.36 m (184 ft. $10\frac{1}{2}$ in.) not only improved the Olympic record by more than a full meter, it bettered Gordien's silver-medal throw by 1.56 m (5 ft. 1 in.). Oerter had clearly cut the gordian knot in his first Olympic effort.

A car wreck in 1957 nearly ended Oerter's athletic career, and his life. He overcame his injuries in time for the 1960 Summer Olympics in Rome, Italy. Gordien's world record had been beaten twice in the four years leading up to the Rome games. The Polish champion Edmund Piatkowski had thrown 59.91 m (196 ft. $6\frac{1}{2}$ in.) in the summer of 1959. A mere fortnight before the opening ceremonies in Rome, the American Richard Babka had tied Piatkowski's world record. During the Olympic trials, Babka had given Oerter his first defeat in two years. In Rome, when Oerter was behind heading into his final throw, Babka gave his teammate some advice about the position of Oerter's left arm during his windup. Oerter heeded the advice and proceeded to break his own Olympic record with a throw of 59.18 m (194 ft. 2 in.). Babka's best throw was more than a meter shorter than Oerter's, and Babka had to settle for the silver medal. I respect Babka for helping his teammate in the purest form of sportsmanship.

In the four years between the Rome games and the 1964 Summer Olympics in Tokyo, Japan, Oerter managed to set a new world record. In each of the years 1962, 1963, and 1964, he threw the discus farther than any man had ever thrown it. His record-setting throw in 1962 established him as the first to throw a discus farther than 200 ft (60.69 m). Oerter's record throw of 62.94 m (206 ft. 6 in.) on April 25, 1964, was eclipsed on August 2 of that year by Ludvik Danek of Czechoslovakia by 1.61 m (5 ft. $3\frac{1}{2}$ in.).

As the world record holder, Danek was the favorite in the Tokyo games, which began on October 10, 1964. The deck was further stacked against

Oerter when he sustained a rib injury just a week before the Olympic competition would get under way. Behind Danek and American teammate David Weill, Oerter was in third place prior to his fifth throw. With grit and determination symbolic of the Olympic ideal, Oerter broke his own Olympic record with a throw of 61.00 m (200 ft. $1\frac{1}{2}$ in.). Oerter was about half a meter better than Danek, who took home the silver.[3] Weill was the bronze medalist. Oerter had once again overcome injury and underdog status to win the gold while establishing a new Olympic record.

The Mexico City games in 1968 proved yet one more setting in which Al Oerter began the competition as the underdog. His American teammate Jay Silvester entered the games as the strong favorite. Silvester held the world record in 1961 and then again in 1968, knocking Ludvik Danek off the top spot. Silvester threw 68.40 m (224 ft. 5 in.) on September 18, 1968, just a few weeks before the start of the Mexico City games. Silvester then broke Oerter's Olympic record in the qualifying round, with a throw of 63.34 m (207 ft. 10 in.). Oerter was once again dogged by injuries, this time to his thigh and neck. On a rainy October 15, three days before Beamon's epic leap, Oerter returned to the summit with a throw of 64.78 m (212 ft. 6 in.), retiring Silvester's day-old Olympic record. Despite his great record-setting qualifying throw, Silvester was unable to medal. The German Lothar Milde earned the silver; Ludvik Danek won the bronze. Silvester's best Olympic finish was his silver medal in 1972.

Oerter had done it! Four golds in four straight Olympics. Each time, he had battled injuries and some of the best discus throwers the world had ever seen. Each time, Oerter set a new Olympic record with his gold-medal throw. Oerter emerged from retirement to compete for an Olympic spot in 1980, finishing fourth at the trials and having to settle for a slot as an alternate.[4] Oerter did set a personal best in 1980 with a throw of 69.46 m (227 ft. $10\frac{3}{4}$ in.). Not bad for a guy nearly 44 years old! Think about that. When he threw that discus in 1980, Oerter was more than twice his age when he first won gold in Melbourne. Suppose the United States had

[3] Danek won the gold in the 1972 Munich games, the first summer games in two decades in which Oerter did not compete.

[4] Oerter would not have been able to attend the Moscow games in 1980 even if he had finished in the top three. The United States boycotted the Moscow games as a protest to the Soviet invasion of Afghanistan in 1979.

competed in Moscow in 1980. Suppose further that one of the regular discus competitors had been unable to compete for some reason. Would you have bet against Al Oerter in a bid to win a fifth gold medal? Not me.

As of this writing, the world record holder in men's discus throw is the German Jürgen Schult with his 1986 throw of 74.08 m (243 ft. $\frac{1}{2}$ in.). Note that the current world record is 4.62 m (15 ft. $1\frac{3}{4}$ in.) farther than Oerter's best throw. Schult did win gold in the 1998 games in Seoul and the silver in the Barcelona games in 1992. But even the top discus thrower of the past twenty years never came close to Oerter's Olympic medal feat. Schult's nearly 22-year reign (again, at the time of this writing) at the top of the mountain, however, is by far the longest reign of any discus thrower. He just never managed four in a row in the Olympics.

The current world record in the women's discus throw is 76.80 m (251 ft. $11\frac{1}{2}$ in.), set by Gabriele Reinsch of Germany in 1988. The fact that the women's record is more than 2 meters farther than the men's reflects the fact that the women's discus (1 kg) is half the mass of the men's discus (2 kg).

The past two decades have been quiet on the record-setting discus front. The 2008 Summer Olympics in Beijing, China, did not see any discus records fall.

The Throw

Legend has it that while he was in high school, a discus landed at the feet of Al Oerter, who was running track at the time. He picked up the discus and threw it back farther than any of the other discus throwers could throw it. The track coach then made Oerter a discus thrower. Smart move! Oerter would go on to set the national high school record for the discus throw.

So, how hard is it to throw a discus? To throw it more than a few meters, that is. Do you think you could pick up a discus lying at your feet and throw it, say, 15 m (nearly 50 ft.)? Some of you athletic types could do this. Most of us, I guess, would struggle to get the discus to a distance that is only about one-fifth of the world record. We certainly won't get much distance if we try to throw it like a baseball. And we certainly won't make any news trying to throw the discus like a Frisbee. It takes

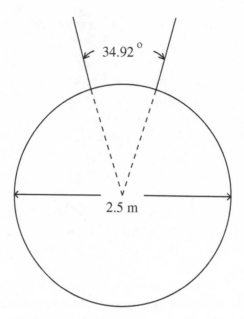

Figure 8.3. The discus throw circle, with arc in which throws must be made.

an enormous amount of strength and skill to throw a discus to some respectable distance.[5]

A discus thrower must constrain his or her throwing motions to a circle of diameter 2.5 m (8 ft. $2\frac{1}{2}$ in.). The discus must leave the thrower's hand within a 34.92° arc.[6] Figure 8.3 shows a rough sketch of the discus-throw circle and the arc within which throws must be made (the arc protects spectators from getting hit!). Oerter had but 4.9 m² (~53 ft.²) to achieve glory.

Before getting into some detailed physics, let's think a little bit about strategy. What do you think is the best way to launch a discus from the circle in Figure 8.3? It probably makes sense to throw it near the front of the circle, since distance measurements are made from the circle's front

[5] A delightful instructional booklet is John Le Masurier's *Discus Throwing* (2nd edition), British Amateur Athletic Board, 1967.

[6] See, for example, the 2008 *Competition Rules* by the International Association of Athletics Federations (IAAF). The rules book may be downloaded at the IAAF's web site, www.iaaf.org.

inside edge.[7] What else could you do? Would you simply stand at the front and heave the discus, or would you want a running start? For the latter option, you would need to begin your motion at the back of the circle. How should your arm be oriented when you release the discus? Should it be out from your body, or should it be cocked so that you throw the discus like a baseball or football? Here we see that sports and physics have something in common—puzzle solving. We have seen already that a physicist will use physical laws and models to explain phenomena. Throwing a discus is similar. You are given a discus and told to throw it while remaining in a circle of specified size. The puzzle to solve is to figure out the best set of body motions for throwing a discus as far as you possibly can. Physics helps us solve the puzzle.

The circle itself is made of a hard surface like concrete. Though it may seem desirable to have a nice cushy surface, think about energy when you visualize such a surface. If you wish to impart as much translational kinetic energy to the discus as then you possibly can, then you want to minimize energy losses during the throwing motion. Translational kinetic energy is defined as

$$KE^{\text{trans}} = \tfrac{1}{2}\, mv^2, \tag{8.1}$$

where v is the speed of an object's center of mass. I discussed rotational kinetic energy in chapter 6, and I mentioned translational kinetic energy in qualitative ways in chapters 5 and 6. I give equation (8.1) now because I want you to realize that maximizing the launch speed of the discus is the same thing as maximizing its translational kinetic energy. Let's think about a soft surface like grass or clay. Now imagine Al Oerter with his 6 ft. 4 in. (1.93 m) height and 280 pound (127 kg mass) frame stepping into the circle. He would leave footprints as he stepped on the soft surface. He would probably leave divots if he were spinning around while he was about to heave the discus. Unlike the springboards that Greg Louganis used, Al Oerter would not be able to recover the energy lost in making footprints and divots. Since concrete's surface does not deform much, it is more "elastic" than soft dirt, at least in the physics sense of the word.

---

[7] Imagine that the circle's center and the landing point of the discus are connected by a line. The point on the front inner edge of the throwing circle that lies on that imagined line is the reference point for the range measurement.

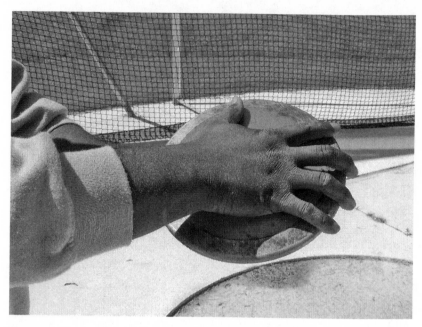

Figure 8.4. Former Lynchburg College discus thrower Rory Lee-Washington holds the discus.

Now that we have the right surface for the circle, let's set Oerter in motion. He begins in the rear of the circle with his back to the throwing area. He holds the discus in his hand much like what you see in Figure 8.4. Oerter drops the discus to arm's length and makes a few casual swings, much like the pendulum in a large grandfather clock. Imagine huge Olympic crowds with a hushed anticipation as the great Al Oerter readies himself for the throw. He knows that he has just the diameter of the circle in which to accelerate the discus to the greatest launch speed he can muster. Oerter also knows what other discus throwers know—that he can accelerate the discus over a distance greater than the circle's diameter. By taking one-and-a-half turns before the throw, he accelerates the discus through a distance about three times the circle's diameter. His movements are designed to release maximal stored energy into the discus's kinetic energy by transferring some of his own kinetic energy to the discus.

At the start, Oerter swings his right arm far behind him so that the discus nearly faces the throwing direction. The muscles and tendons in his powerful right arm are stretched, thus storing potential energy. His left arm is stretched out in front of his body. Oerter begins turning counterclockwise (as seen from above). As he turns, he drops his center of mass by bending his knees. Some gravitational potential energy is consequently transferred into stored energy in his stretched leg muscles. During the turn, Oerter simultaneously moves his center of mass toward the front of the circle. It is during the initial turn that Oerter begins what amounts to a small "run" across the circle. The final full turn allows him to release a good fraction of the energy stored in his stretched muscles as his arm comes forward and his legs straighten out. His arm then whips the discus out much in the fashion of a medieval trebuchet, or war machine. At the point of release, Oerter's fingers are grasping the back edge of the discus. Their line of force is nearly through the discus's center.

The aforementioned line of force cannot be exactly through the center because Oerter wants the discus to spin as it leaves his hand. The discus will leave the right-handed Oerter spinning clockwise (as viewed from above). By now, you may know why he wants the discus to spin. From our discussion of angular momentum conservation, we know that it takes a torque to change the direction of the discus's angular velocity vector. Just like the rifling that initiates spin in a bullet, spinning a discus gives it stability. Air friction will impart a small torque to the discus, but not enough to alter greatly the discus's orientation. Stability means that the discus will mostly maintain its initial knife-blade orientation with the air, keeping drag forces relatively low.

To get an estimate for the spin rate, imagine rolling the discus along the ground so that its center of mass speed is its launch speed. Take the speed to be 25 m/s. If the discus rolls without slipping, its center of mass speed, v, will be related to its angular speed, ω, by

$$v = r\omega, \tag{8.2}$$

where r is the radius of the discus. Put the numbers in, solve for ω, and get

$$\omega = \frac{v}{r} \simeq \frac{25\,\text{m/s}}{0.11\,\text{m}} \simeq 227\,\text{rad/s} \simeq 36\,\text{rev/s}, \tag{8.3}$$

where in the last step I have used the fact that there are 2π radians in one revolution. The greater the rotational rate, the larger the torque needed to move the angular velocity vector by a given amount. But Oerter cannot use most of the force from his fingers and hand to spin the discus, because he needs a substantial component of that force along the center of the discus to give it a large translational speed. Oerter must thus find a balance between large spin rate and large translational speed. He chooses just enough rotational speed for reasonably good stability while maintaining a sizable force along the center of the discus. I realize that that last sentence is not terribly quantitative! It is trial and error, during lots of practice, coupled with what the body is physically capable of doing, that decide the ideal balance, a balance better than a scientist can conjure. Observations suggest a spin rate about one-fourth of what I calculated in equation (8.3).

Where in his whipping motion should Oerter release this discus? If something constrains an object to move on a circular path, what happens when that constraint is removed? Check out Figure 8.5, which shows a popular conceptual question from introductory physics. A circular tube sits on a smooth horizontal table. A marble has been rolling through the

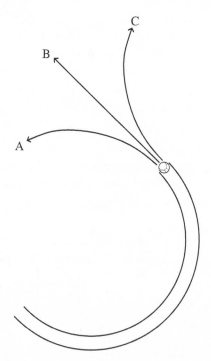

Figure 8.5. Overhead view of a marble rolling out of a circular tube. Which path does it follow after leaving the tube?

tube. You see an overhead view of the marble about to exit the tube. Which path does it follow? Does it follow path A, continuing along its circular path? Could it follow path B and move along a straight line? Or, does it feel an outward force and move along something like path C? If you said "B," give yourself an A+! While moving through the tube, the marble experiences an upward force from the table that balances the marble's weight. We know from Newton's first law that an object moving with constant velocity (i.e. constant speed *and* constant direction) will continue to do so unless acted upon by a net external force. Even if the marble's speed is constant in the tube, the direction of its velocity certainly changes. The tube's wall must provide a contact force on the marble that causes its velocity to constantly change direction. But which part of the tube exerts that force, the inner part or the outer part? Think about the direction of the *change* in velocity. The marble's velocity vector keeps getting turned to the left as it traverses the tube. That means that the outer part of the tube must exert the force, meaning that the force is *inward*, toward the center of the circular tube, though the marble never loses contact with the outer wall of the tube.

If the conclusion in the previous paragraph still seems a little fuzzy, think about sitting in the right side of a car with your right shoulder almost in contact with the car door. The car makes a quick left turn, much like that at the Indy 500, where you can make only a left turn. You feel the right side of the car pushing on you, right? Newton's first law tells you that you want to keep moving along a straight line, but the car wants to turn left. You thus need a force from the car pointing toward your left to change your velocity vector. The sensation you have that you are "thrown outward" fools you. There is nothing throwing you to your right. You simply want to keep moving along a straight line, as Newton's first law tells you to do, when all of a sudden a net external force—from the car—comes along and shoves you to the left.[8] Just before releasing

---

[8] In my example with you in the car, I apply Newton's laws from the point of view of a stationary observer *outside* the car. I have no problem explaining your acceleration out there. Inside the car, however, is a different story, because that represents an accelerated reference frame. Newton's first law is not valid in such a "non-inertial" frame (inertial frames are ones in which Newton's laws hold true). When studying circular motion, you must be very careful to analyze the problem in an inertial frame of reference (or as close to one as we can have on a rotating Earth).

the discus, Oerter moves it along an arc. He releases it as his hand passes near the front edge of the circle.

We call the net inward force a *centripetal* force (centripetal means "center seeking"). At the instant a particle moves with speed v along the arc of a circle of radius r, it has an acceleration component toward the center of the circular arc with magnitude

$$a_c = \frac{v^2}{r}. \tag{8.4}$$

We can estimate the distance from Al Oerter's axis of rotation near his spine to the center of the discus to be about a meter. To launch at 25 m/s requires an acceleration of about

$$a_c \simeq \frac{(25\,\text{m/s})^2}{1\,\text{m}} = 625\,\text{m/s}^2 \simeq 64\,g. \tag{8.5}$$

Newton's second law, given by equation (3.6), tells us how to obtain the force if we know the acceleration. The centripetal force is roughly the discus mass of 2 kg times the acceleration found above. Plugging in the numbers gives

$$F_c \simeq (2\,\text{kg})\,(625\,\text{m/s}^2) = 1250\,\text{N} \simeq 281\,\text{lbs.} \tag{8.6}$$

Tack on another 4.4 lbs. for the weight of the discus.[9] Oerter had to exert a force with his right arm comparable to his own weight! You can see why discus throwers need an enormous amount of strength conditioning before being able to compete at the highest level.

One other thing to note here is Oerter's advantage of being tall. Equation (8.4) tells us that the centripetal acceleration goes down as the radius goes up, meaning that the needed inward force goes down. Put another way, if two athletes are capable of exerting the same inward force, they are capable of giving a discus the same centripetal acceleration. But the athlete with the longer arms, meaning a larger r in equation (8.4), is able to give the discus a larger launch speed. Who said life is fair?! If you have

---

[9] Remember that forces must add as *vectors*. The centripetal force and the force needed to hold the discus are not in the same direction. Because the weight of the discus is small compared to the centripetal force, I can ignore the weight in my discussion.

the genes to grow tall, you have a leg up in sports like basketball and discus throwing (other sports, too!). You still need to invest a lot of blood, sweat, and tears to be the best, though.

If the point in the previous paragraph is not crystal clear, perhaps a return to the car analogy for a moment will help. If you take a curve too fast, meaning that the frictional force from the road on your tires is not great enough to keep you moving in a circle of a given radius at the speed you want, what happens? You may answer that your car is "thrown outward," but that's not what really happens. Friction provides a certain centripetal force, thus fixing the centripetal acceleration. If a_c is fixed in equation (8.4), there corresponds a particular r for a given v. The faster you go, the larger the radius you need. In other words, the road cannot pull your car inward enough. Your car is not thrown outward; rather, it is not thrown inward enough. The reduced speed limit signs you see at the start of sharp curves are there for a reason!

My goal here is to give you a qualitative idea of how Oerter was able to transfer energy from his body to the discus. I have *not* tried to offer a definitive account of the sequence of steps one must go through to throw a discus. If you want to read more details from a coach's point of view, read Le Masurier's fine booklet (see footnote 5 above).

Flight in Vacuum

Now that Oerter has released the discus, we need to examine the aerodynamics of its flight. As I have done throughout this book, I will first consider the flight of the discus in vacuum. For that, I once again employ the constant-acceleration projectile equations (3.7) and (3.8) from chapter 3. Suppose Oerter launches the discus from a height h above the ground, meaning $y_0 = h$ in equation (3.8). After a time T, the discus hits the ground at $y = 0$. If the discus leaves Oerter's hand with a center of mass speed v_0, and its velocity vector is at an angle θ_0 to the ground, then $v_{0x} = v_0 \cos\theta_0$ and $v_{0y} = v_0 \sin\theta_0$. A little algebra reveals that the range of the discus is

$$R = \left(\frac{v_0^2 \cos\theta_0}{g} \right) \left(\sin\theta_0 + \sqrt{\sin^2\theta_0 + \frac{2gh}{v_0^2}} \right). \qquad (8.7)$$

That equation may seem a little cumbersome, but we can simplify things a tad if we optimize Oerter's throw. Suppose Oerter's strength and technique maximize v_0; he simply cannot throw the discus any faster. What launch angle, $\tilde{\theta}_0$, will maximize the range R? The answer turns out to be rather simple.[10] The optimum angle may be determined from

$$\sin^2 \tilde{\theta}_0 = \frac{1}{2\left(1 + gh/v_0^2\right)}. \tag{8.8}$$

This equation provides for a relationship between the optimum launch angle and the launch speed. We can thus eliminate one of them in an expression for the maximum range, R_{max}. If we use equation (8.8) for $\tilde{\theta}_0$, equation (8.7) becomes

$$R_{max} = \frac{v_0^2}{g} \sqrt{1 + \frac{2gh}{v_0^2}}. \tag{8.9}$$

That's better than equation (8.7), right? We can get even simpler if we solve equation (8.8) for v_0 in terms of $\tilde{\theta}_0$ and plug that into equation (8.7). After some fun algebra that I'll leave for you, we get

$$R_{max} = h \tan\left(2\,\tilde{\theta}_0\right). \tag{8.10}$$

Now that's as simple as I can make it! If you know the launch speed, use equation (8.9) to obtain the maximum possible range. If it's easier to measure the launch angle, then use equation (8.10).[11]

I'll stop spewing out equations now and get to some numbers. Oerter stood about 1.93 m (6 ft. 4 in.) tall. Suppose he releases the discus from a height of 1.8 m (~5.9 ft.) with launch speed 25 m/s (~56 mph). Equation (8.8) predicts an optimum release angle of about 44.2°; equation (8.9) gives roughly 65.6 m (~215 ft.) for the maximum possible range.

There are a couple of things to notice about the numbers we just calculated. The first is that the optimum launch angle is not that far removed

---

[10] For those of you fluent in calculus, you know that the word "maximize" means you need to take a derivative. If you wish to derive equation (8.8), evaluate $dR/d\theta_0$ at $\theta_0 = \tilde{\theta}_0$ and set that expression to zero. You then get to do some fun algebra!

[11] Equations (8.7) to (8.10) rely on the fact that the initial and final heights differ. Though deriving those is straightforward, and I have seen them in several books and papers, I first saw these results in "Maximizing the range of the shot put," D. B. Lichtenberg and J. G. Wills, *American Journal of Physics* 46 (5), 546–49 (1978).

from 45°, which is the optimum launch angle if the initial and final heights are the same. The reason is that the difference between the initial and final heights is small compared to the range of the discus. Had Oerter been able to throw from the ground with a launch angle of 45°, the time of flight would have been only about 1.4% shorter than if he had released it at 1.8 m off the ground.

The second thing to notice from the numbers we computed, and perhaps more interesting than the first, is that the predicted range is only about 5.6% removed from Oerter's personal best. That's not too bad, considering that we have no air around our discus! But this result may seem contradictory, since vacuum results are not always great compared to what actually happens in the real world. One makes, for example, something like a factor of 2 error if trying to predict the range of a hit baseball with no air around. You can quibble with my choice of launch speed, but that's actually a pretty good estimate for what a top-notch discus thrower like Al Oerter could achieve. If we know that the air will reduce the range of the discus, there must be something else that helps *add* to the discus during its flight. We need to turn our attention to what it is that "lifts" the discus.

Flight in Air

Putting air into our problem means that we need once again to determine all the forces on the discus. By now you know that there must be a gravitational force from the Earth. That was used implicitly in the preceding discussion. Since a men's discus has a mass of 2 kg, the gravitational force, or weight, is

$$F_g = m\,g = (2\,\text{kg})\,(9.8\,\text{m/s}^2) = 19.6\,\text{N} \simeq 4.41\,\text{lbs.} \qquad (8.11)$$

A women's discus has half the mass, so its weight is 9.8 N (~2.20 lbs.).

We have met the drag force a few times already, and the discus certainly feels air resistance during its flight. Let's once again employ the air drag equation, given by equation (4.2) as

$$F_D = \tfrac{1}{2}\,C_D\,\rho\,A\,v^2. \qquad (8.12)$$

With the exception of dividing up cycling into a couple of different regimes for body orientation, I have used constant values for the dimensionless drag coefficient, C_D. I showed you experimental data for a soccer ball's drag coefficient in the last chapter; but I still used a constant C_D because my purpose there was to show how you could count good kicks and bad kicks. In this chapter, I wish to model the flight of the discus with the goal of predicting a range that is close to Oerter's actual range. That goal cannot be achieved without a varying C_D.

Unlike a soccer ball, which always presents the same cross-sectional area to the air, a discus presents a range of cross-sectional areas. The men's discus has a diameter of 22 cm (~8.66 in.); the diameter of the women's discus is 18 cm (~7.09 in.). If the discus could be thrown such that it looks like a sail, with its greatest cross-sectional area exposed to the air, then $A \simeq 0.038$ m^2 (~0.41 ft.^{2}) for men and $A \simeq 0.025$ m^2 (~0.27 ft.^{2}) for women. The aforementioned "sail" orientation has the discus facing the air as in Figure 8.6. If the discus passes through the air in the orientation shown in Figure 8.7, the cross-sectional area facing the air is minimized. I estimate that the area in Figure 8.7 is about 16% (or roughly one-sixth) of the area in Figure 8.6.[12] The question is, which A should be used in equation (8.12)? As we did with Lance Armstrong on his bike, we will account for variations in cross-sectional area through use of the drag coefficient. For that, we need to discuss what the discus looks like from the air's point of view.

Consider Figure 8.8. The center of mass of the discus moves with velocity $\vec{v}_d$. We also allow for the presence of wind. For simplicity, let's keep the wind parallel to the ground, and in the plane formed by the initial launch velocity and a vector perpendicular to the ground. The wind velocity vector $\vec{v}_w$ in Figure 8.8 implies that the wind is in Oerter's face if we imagine that he throws from left to right in that figure. When thinking about what speed we need to use in equation (8.12), consider it from the point of view of the discus. Forgive me if I anthropomorphize the discus, but it feels air hitting it because it *moves through* air *and* because there is also wind, which the discus would feel even if it weren't moving. What we need

---

[12] To obtain that percentage, I estimated from Figure 8.7 that the end view of the discus consists of a central rectangle with a triangle on each end. I used distance measurements from page 43 of the NCAA's *Cross Country and Track & Field: 2008 Men's and Women's Rules*. The ratio of areas is then that total end area divided by the circular area of the "sail" orientation in Figure 8.6.

Figure 8.6. This top (or bottom) view of the discus shows the maximum cross-sectional area. If oriented this way, the discus might look to oncoming air like a sail.

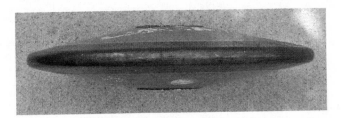

Figure 8.7. This orientation of the discus presents the smallest cross-sectional area to the air.

therefore is the *relative* velocity of the discus with respect to the air. Think back to Figure 2.6, and then note that Figure 8.8 shows that $\vec{v}_r = \vec{v}_d - \vec{v}_w$. We thus need to use the relative speed, v_r, for v in equation (8.12). The direction of the drag force $\vec{F}_D$ is opposite the vector $\vec{v}_r$.

Recall that I introduced the Magnus force in the preceding chapter. The idea there was that a spinning soccer ball that whips air off in a

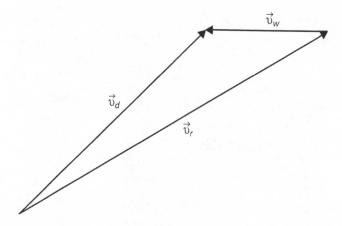

Figure 8.8. The relationship between the velocity of the discus, $\vec{v}_d$, the velocity of the wind, $\vec{v}_w$, and the velocity of the discus relative to the air, $\vec{v}_r$.

direction not aligned with the direction of motion feels a Magnus force perpendicular to that direction of motion. I mentioned in footnote 10 in Chapter 7 that the expression for the Magnus force I used in modeling the flight of a soccer ball derives from a more general expression for the lift force. Although the discus does whip air off to one side because of its spin, and thus feels a Magnus force, I wish to ignore that effect here. Instead I want to focus on the lift the discus feels because of its shape and orientation with respect to its velocity relative to the air. The Magnus force causes the discus to curve slightly in its flight, meaning that the trajectory, like the soccer ball's, is three-dimensional. By considering only the lift force due to the airplane-wing-like shape of the discus, I keep the problem two-dimensional.[13]

When I first discussed air drag, back in chapter 4, I asked you to imagine holding your arm out your car window. You feel a small force

[13] For a three-dimensional analysis, see "Optimal discus trajectories," Mont Hubbard and Kuangyou B. Cheng, *Journal of Biomechanics* 40, 3650–3659 (2007). An older article is "The Dynamics of Discus Throw," T. C. Soong, *Journal of Applied Mechanics* 43, 531–36 (1976). If you wish to pursue discus modeling beyond what I do here, be warned that keeping track of all the angles can make your head spin as rapidly as the discus!

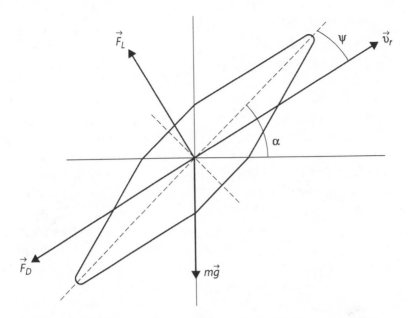

Figure 8.9. The angles needed to model the flight of a discus through air are the angle of attack, ψ, and the angle, α, between the horizontal ground and the long axis of the discus. Also shown are the forces on the discus: weight ($m\vec{g}$), drag ($\vec{F}_D$), and lift ($\vec{F}_L$).

with your palm facing the ground and a large force with your palm facing the wind. We understood that principle from the idea that a larger cross-sectional area exposed to the wind makes for a greater drag force. Think now about what happens if your hand is oriented between the two aforementioned extremes. In other words, suppose you are in the passenger seat with your arm sticking out of the window. Start with your palm down and then rotate your hand 45° back so that air hits your palm with a glancing blow. What direction of wind force do you feel now? You probably feel as if the wind is pushing your hand backward and upward. You feel not only a component of wind force antiparallel to the direction of motion (backward), but you feel a perpendicular (upward) component as well. Your hand thus feels a lift force, just as an airplane wing does.

Now look at Figure 8.9. You see a sketch of a discus while at some point in its trajectory. The velocity of the discus relative to the air, $\vec{v}_r$, is the only velocity vector I need to show. The drag force, $\vec{F}_D$, is, as I have already stated, antiparallel to $\vec{v}_r$. Notice in the figure that air will hit the bottom of the discus. The discus must then deflect the air downward. Newton's third law tells us that the air must push the discus upward. You thus see a lift force, $\vec{F}_L$, pushing the discus upward and perpendicular to $\vec{v}_r$. Notice that there will be no lift force if the angle ψ, called the "angle of attack," is zero. In that case, air moves around the top and bottom of the discus equally, and there is no deflection of air away from the long axis of the discus. Note also that if the long axis of the discus were *below* $\vec{v}_r$, the lift force would be down and to the right, i.e. rotated 180° from its orientation in Figure 8.9. That would be analogous to sticking your arm out the passenger window with your palm down and then rotating your hand 45° such that the wind hits the back of your hand. You would then feel the wind pushing your hand backward and downward. In that case, we might call the effect "negative lift" since the air pushes downward instead of lifting upward.

To model the lift force, I use the equation I introduced in footnote 10 in chapter 7, namely,

$$F_L = \tfrac{1}{2}\, C_L\, \rho\, A\, v^2. \tag{8.13}$$

Note the appearance of ρ, the air density. You probably know that air density decreases with increasing elevation. Equation (8.13) tells us why a helicopter can go only so high. There is a certain elevation at which maximum lift exactly balances a helicopter's weight. Any higher, ρ decreases such that F_L can no longer match a helicopter's weight.

As with the drag equation, it is not obvious what A should be. We can, however, account for variations in the cross-sectional area with the dimensionless lift coefficient, C_L. We take A in equations (8.12) and (8.13) to be the maximum cross-sectional area that could be exposed to the air, i.e. $A \simeq 0.038\,\mathrm{m^2}$ ($\sim 0.41\,\mathrm{ft.^2}$) for a men's discus and $A \simeq 0.025\,\mathrm{m^2}$ ($\sim 0.27\,\mathrm{ft.^2}$) for a women's discus. The dimensionless coefficients C_D and C_L are then taken to be functions of the attack angle, ψ. I assume that the angle the long axis of the discus makes with the horizontal, α in Figure 8.9, remains fixed. That is true as long as we ignore any net external torques on the discus from the air. Oerter sets α upon releasing the discus. The attack angle, ψ, changes throughout the flight of the discus because $\vec{v}_r$ continually changes direction.

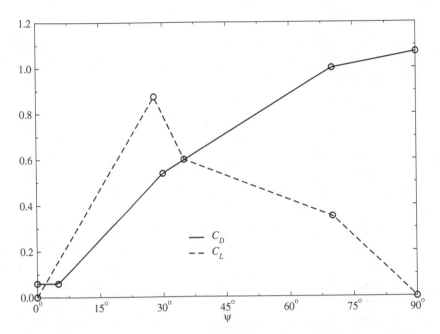

Figure 8.10. Data for C_D and C_L as functions of the attack angle, ψ. Linear interpolation is used between the data points to make continuous functions. (Data courtesy of Cliff Frohlich)

Figure 8.10 shows data for C_D and C_L as functions of ψ.[14] The figure shows only five experimental data points for each of C_D and C_L; I have simply drawn straight lines between the data points. Though the "real" values of C_D and C_L at all possible values of ψ are surely different from what I show in Figure 8.10, the right orders of magnitude and correct qualitative physics are contained in the plots. We expect the drag coefficient to increase steadily from a small value at zero attack angle (palm facing down in my arm-out-the-window analogy) to a maximum value at 90° (palm directly facing the wind). There is no lift for the two configurations I just mentioned because there the air streaming around all sides of the discus is essentially symmetric.

[14] I thank Cliff Frohlich for graciously allowing me to use data he used in his paper, "Aerodynamic effects on discus flight," *American Journal of Physics* 49 (12), pp. 1125–32 (1981). I first learned of techniques for modeling the flight of a discus from Professor Frohlich's paper, and I am grateful that he allowed me to use some of those techniques in this chapter.

I now do as I have done throughout this book. I throw all the force expressions into Newton's second law, dump everything into a computer, turn the crank on the side of the computer, and wait for the trajectory of the discus to pop out. Those details need not appear here; I'll move on to the results in the following discussion.

Optimum Trajectory

For all of the upcoming results, I'll stick with the same initial height and launch speed I used for the vacuum case: $h = 1.8$ m (~ 5.9 ft.) and $v_0 = 25$ m/s (~ 56 mph). Think about all the aspects of training that Oerter had to go through to be the world's best discus thrower. The limits of his strength conditioning and technique in the discus circle put a corresponding limit on the launch speed of the discus. His training must also allow him to find the launch angle, θ_0, and discus orientation that maximize the range. We saw in Figure 8.10 that a small attack angle means a small amount of drag. But an attack angle around 30° means a lot of lift, which keeps the discus in the air longer. There is thus a battle between drag and lift, a battle those of you familiar with flying know all too well. An airplane would experience no drag in vacuum, but it couldn't fly because it needs air moving around its wings to generate lift. Oerter cannot control the attack angle; he *can* control the angle at which the discus is oriented with respect to the ground. That is angle α in Figure 8.9. After fixing h, and v_0, Oerter's training had to include finding the values of θ_0 and α that yield maximum range. Even if a nerdy physicist like me is able to find those special angles that optimize the trajectory, it is by no means an easy task to turn theory into reality. Getting one's arm and hand just right for the perfect release takes years of training and practice.

To find the optimum trajectory, I need to solve Newton's second law equation using particular values of θ_0 and α. That will give me a value for the range. I then vary θ_0 and α a little and find a new range. I keep doing this over and over until I find values of θ_0 and α that give the maximum range. I further complicate things by allowing the wind speed, v_w, to vary. There is nothing in the rulebook about wind speed when it comes to discus throwing. Anything goes. With so many parameters varying, you can see why a computer is such a wonderful tool.

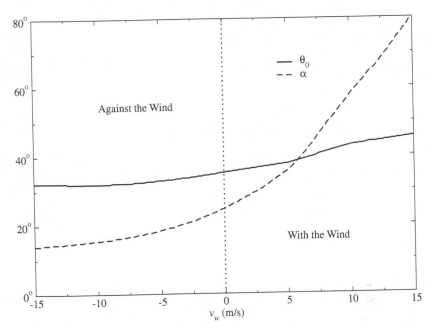

Figure 8.11. Values of launch angle, θ_0, and discus orientation, α, as functions of wind speed that give maximum range. Note that $5\,\text{m/s} \simeq 11.2\,\text{mph}$.

Figure 8.11 shows the values of θ_0 and α needed to maximize the range for a given wind speed. A wind speed of $15\,\text{m/s}$ (about 34 mph) is likely to be much greater than what would occur at a discus competition. (I go out that far merely to show trends.) Much else can be gleaned from Figure 8.11. If the wind is in Oerter's face as he throws, he needs to launch the discus with its nose oriented about $10°$ to $15°$ below the launch angle. Given the unlikelihood of wind speeds much greater than about $5\,\text{m/s}$, the strategy when throwing into the wind is to orient the discus about $10°$ below the launch angle. As Figure 8.9 shows, the lift force at the very beginning of the flight will have a downward component. Once the velocity vector of the discus relative to the wind drops below the nose of the discus, the lift force has an upward component. The lift will thus be upward during the majority of the flight.

Now look at Figure 8.11 again and note what happens if Oerter releases the discus with the wind hitting his back. As the wind speed gets greater, the inclination angle of the discus needs to get greater. Beyond $10\,\text{m/s}$ or

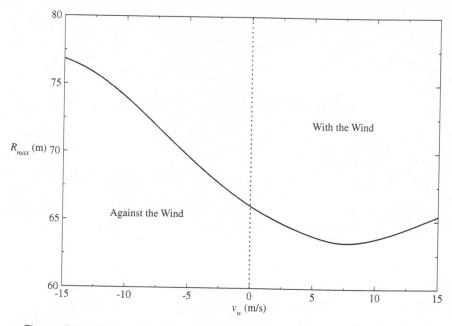

Figure 8.12. Maximum discus range as a function of wind speed. At each v_w, angles from Figure 8.10 are used. Note that 70 m $\simeq$ 230 ft.

so, α seems almost unphysically large. At that point, the discus must look like a sail if it is to achieve a large range. The big idea here is that if the wind is in your face, you need to angle the discus almost as if you are skipping stones on a pond. With the wind at your back, you need to loft the discus up like a sail.

Figure 8.12 shows the maximum range for each wind speed. For each value of v_w, I use the corresponding values of θ_0 and α from Figure 8.11. That graph may at first contradict your common sense! With running and jumping events, like Beamon's long jump, the rules specify maximum values of allowable wind speed. Beamon is helped by a strong wind on his back. What Figure 8.12 tells us is that if Oerter throws *into* the wind he gets a *greater* range than if he throws with no wind or with the wind at his back. Wind blowing in the direction of the throw simply doesn't help very much. Think about equation (8.13). The speed we need in that equation is the speed of the discus *relative* to the air. When do you get a large relative speed? When you throw *against* the wind. Another way

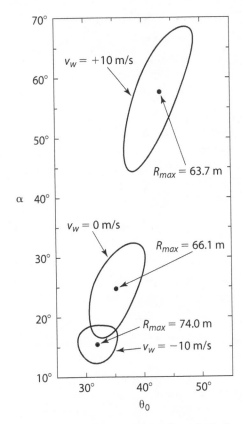

Figure 8.13. Discus inclination α versus launch angle θ_0 for three wind cases. Each "island" encompasses values of θ_0 and α that give a range within 1 m of the maximum range.

to think about it is to imagine the air moving over the discus, much like air moving over an airplane wing. If the relative air speed is small, there won't be much lift. Don't hold your breath waiting for a parked airplane to lift into the air when a breeze flows over its wings! An airplane needs to accelerate along a runway to get a speed great enough relative to the air to generate enough lift to get it off the ground. The lift on a discus works in a similar way. The greater the speed of the air moving across its surface, the greater the lift on the discus.

Where do you think the margin of error is smallest—throwing into the wind, throwing with no wind, or throwing with the wind? Take a look at Figure 8.13 (which is not a map of three of the Hawaiian Islands!). What you see there is a plot of discus inclination versus launch angle for the three aforementioned cases: throwing into the wind ($v_w = -10$ m/s), throwing with no wind ($v_w = 0$ m/s), and throwing with the wind ($v_w = +10$ m/s). The dot located roughly in the middle of each "island" represents the (θ_0, α) point that gives the maximum range for a given wind speed. The "island" encompasses surrounding (θ_0, α) points, all of which give a range within 1 m of the maximum range. Note that while throwing into the wind gives the greatest range, it also has the smallest margin of error. Here we see another example in nature in which there is no free lunch! The area of the "with wind island" is almost 50% larger than the "no wind island" and more than four times bigger than the "against wind island." If you are a discus thrower and you are lucky enough to enter the launch circle with a strong wind blowing opposite the direction you wish to throw, you are going to need a technique of highest quality if you want to set any records.

I'll finish this chapter with a graph of what one of Oerter's throws might have looked like. Suppose there is no wind ($v_w = 0$) and Oerter unleashes the perfect throw. The discus leaves his hand 1.8 m off the ground at a speed of 25 m/s. From Figure 8.11, we determine that $\theta_0 \simeq 35.25°$ and $\alpha \simeq 24.75°$. Throw all that into the computer and you get Figure 8.14. The range is 66.1 m ($\sim$217 ft.), nearly identical to the 65.6 m ($\sim$215 ft.) range found earlier when we optimized the vacuum throw. What is not so similar is the release angle. For vacuum, $\theta_0 \simeq 44.2°$ gave the maximum range, about 9° greater than the result in air. The simple vacuum results give about the same range and time of flight as the results in air, but the trajectory predicted from vacuum physics is wrong. As you can see in Figure 8.14, the optimum vacuum trajectory goes about 4 m ($\sim$13 ft.) higher than the trajectory in air. The smaller launch angle in air allows the discus to "cut" through the air and make positive use of the lift force over much of its trajectory. We thus have the remarkable result that with no wind, the loss in range due to the drag force is almost completely compensated by the gain in range due to the lift force. Throwing into the wind allows one to use lift even more, and to exceed what one could do in a vacuum. So you see, air resistance isn't always such a bad thing!

In his post-discus-throwing life, Al Oerter turned his energies to art. He created paintings from the spatter that resulted from a discus hitting paint. Oerter played a significant role in creating *Art of the Olympians*,

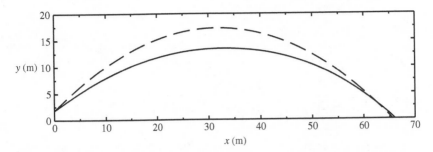

Figure 8.14. The optimum "no wind" trajectory in air (solid line). The range is roughly 66.1 m (~217 ft.). The no wind time of flight is about 3.71 s. The vacuum (dashed line) range is roughly 65.6 m (~215 ft.). The vacuum time of flight is about 3.66 s.

an organization that has assembled artwork from great Olympic athletes.[15] Fellow discus-thrower Rink Babka and chapter 5's subject, Bob Beamon, are among the artists. Let's now leave a giant of the Olympics for a different kind of giant.

[15] Check out the web site www.artoftheolympians.com for more information.

9

It Takes More Than a Big Gut

Caloric Consumption and Linear Momentum

Tradition Like No Other

I'm not referring to the *The Masters* here.[1] I'm thinking about something *much* older. The United States celebrated its bicentennial during my lifetime. I grew up with the big three American sports: baseball, basketball, and football. I learned much about the traditions in each of those sports. But sports traditions here are less than a couple of centuries old. The Japanese have a sports tradition with roots going back a couple of *millennia*: sumo. About the time (1607) that Jamestown, Virginia, became the first permanent English settlement in what would become the United States, sumo came into its own as a Japanese spectator sport. Of all the chapters in this book, this chapter has the least connection to the American sports scene. That means only that it proved to be the most fun for me to research and write. And as mysterious as sumo may be to those in the West, its participants are just as much constrained by the laws of physics as any other group of athletes.

The Nihon Sumo Kyokai, or Japan Sumo Association,[2] is the body that organizes the sport in Japan. Every other month, the country is treated to a *honbasho*, or main tournament. That's six big sumo events every year! Three take place in Tokyo; the other three occur in Osaka, Nagoya, and

---

[1] CBS uses the "Tradition Like No Other" reference in its coverage of *The Masters*, a major golf tournament that began in 1934.

[2] Visit their web site at www.sumo.or.jp.

Fukuoka. For fifteen days, the main tournaments showcase the spectrum of sumo talent. Junior *rikishi*[3] compete roughly every other day in the early sessions for a total of seven bouts per junior *rikishi* for each main tournament. Later prime-time sessions are reserved for the top senior competitors. Top *rikishi* have just one match per day, and it all comes down to wins and losses. If a *rikishi* wins more than 50% of his matches, he is in the *kachikoshi*; a sub 50% record gets one relegated to the *makekoshi*. How one does in a given main tournament influences where one is ranked for the next big event.

Upon seeing a sumo match for the first time, one may be surprised by the amount of time devoted to rituals, as opposed to fighting, especially in matches involving senior *rikishi*. A ceremony known as *dohyo-iri* transpires in the ring before all the big *rikishi* compete; they then leave and return individually as their respective match times approach. Once a top *rikishi* enters the *dohyo*, or circular ring, he engages in rituals rooted in the Shinto religion. Spectators observe clapping and exaggerated stomping (the *shiko* exercise). Just before assuming their fight positions, *rikishi* toss salt into the ring, a Shinto gesture of purification. Each *Rikishi* then takes a position behind his respective *shikiri-sen*, which is a white line behind which a *rikishi* crouches in anticipation of the start of the bout. (See Figure 9.1 for a simple sketch of the *dohyo* and *shikiri-sen*.) The *gyoji*, or referee, gives a signal informing the combatants that they may commence. Do they charge forward? No! There is no gun shot that tells them to charge; there is no bell to signify the beginning of the battle. Each *rikishi* stares the other one down for a little while in a ritual called *niramai*. The glower from a sumo giant is designed to break the will of his opponent. It's psychology time here, not physics time. *Rikishi* may even go back to their respective corners for more concentration. They then get back to their lines and try again. Fans may have to wait the length of time it took Roger Bannister to run the mile back on May 6, 1954—four minutes! Once the *rikishi* come to silent agreement, they charge simultaneously in what is called the *tachi-ai*.

---

[3] An easier word for English-speaking people to use is *sumotori*, which means "one who does sumo." *Rikishi*, which means "strong man," is the preferred term in Japan, so I'll use it here. Whatever term you choose, please do not use the term "wrestler." Sumo is quite different from wrestling.

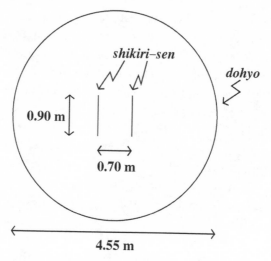

Figure 9.1. Simple overhead sketch of the *dohyo* (ring) and *shikiri-sen* (starting lines). Not shown are all the ritualistic items placed just outside the ring.

But don't blink! The match may be over in a few seconds, and rarely does a match make it to the one-minute mark. The winner is the one who either drives his opponent out of the ring or forces him to touch the fighting surface with something other than a foot. The initial charge is crucial to success. National glory is mere seconds away for the better *rikishi*.[4]

A Name Like No Other

Think of your favorite sport. Is there one name that even a person who has never watched the sport would know? If you love baseball, what about Babe Ruth? In Japan, the baseball name everyone knows is Sadaharu Oh. What about boxing? Most people probably know Muhammad Ali, even if many never saw him fight. There is probably one name that could be

---

[4] I have neither the space nor the scope in this book to give a complete and detailed account of sumo, its various rituals, or its history. I refer you to books devoted to sumo, books written by those with far greater knowledge of sumo than I. A good example is Lora Sharnoff's *Grand Sumo: The Living Sport and Tradition* (Weatherhill, Inc., New York, 1993).

shouted in, for example, Shibuya[5] and everyone would think "sumo." That name is Taiho Koki, or Taiho for short. Born May 29, 1940, Taiho stood 1.87 m (6 ft. $1\frac{1}{2}$ in.) tall and had a mass of about 150 kg (nearly 331 pounds of weight) in his heyday. Taiho's big debut came in September of 1956 at an age when many in the United States go for their driver's licenses. Rising very quickly through the sumo ranks, Taiho was promoted to the 48th *yokozuna* in September 1961, the same month my favorite pro football player, Dan Marino, was born. *Yokozuna* may be translated to "grand champion." It is the highest rank one can attain in the sport of sumo. Once at the summit, one can never be demoted. At any given time, there are only one to four or so *yokozuna*.[6] At 21 years of age, Taiho was the youngest *yokozuna* of all time.[7] After a decade of dominance, Taiho retired in May 1971 with a top-division[8] record of 746 wins and 144 losses. That's a winning percentage of about 84%. Taiho still holds the record for the most *honbasho* wins with 32. There were 69 *honbasho* between the Tokyo event in January 1960 and the Tokyo event in May 1971. By winning 32 of them, Taiho had won over 46% of the *honbasho* in which he competed. In eight of them, he was a perfect 15-0. Now that's dominance!

One match in particular, on Sunday, May 8, 1960, more than a year before his promotion to *yokozuna*, made the world of sumo take notice of Taiho. His first match of that Tokyo *honbasho* was against the mighty Asashio Taro (Asashio for short), the 46th *yokozuna*. With fans screaming in the background, Taiho sprang from the *tachi-ai*, and then managed to get his right arm under Asashio's left shoulder. With enormous strength, Taiho turned Asashio to Taiho's left and forced him out of the ring with a classic *yorikiri* move. *Yorikiri*, which means "drive out," is the most

---

[5] This famous part of the Shibuya Ward in Tokyo is the Times Square of Japan, with the perpetually busy Shibuya Station in the heart of the district. If you love an adrenaline-filled nightlife, I recommend that you visit Shibuya if you are ever in Tokyo.

[6] At the time of this writing, there are two active *yokozuna*. Asashoryu Akinori became the 68th *yokozuna*, and the first from Mongolia, in January 2003. Hakuho Sho became the 69th *yokozuna*, the second from Mongolia, in May 2007.

[7] Kitanoumi Toshimitsu subsequently became the youngest *yokozuna* when he was made the 55th member of the exclusive club in July 1974. Kitanoumi was a mere one month younger than Taiho was at the time of Taiho's promotion.

[8] There are six divisions. The top one is called *makunouchi*, which can have at most 40 members.

popular winning technique. In the span of about five seconds, Taiho had defeated a *yokozuna*. The victory earned Taiho his only *kinboshi*, which means "gold star" and is given when an upper-level *rikishi* knocks off a *yokozuna*.

There are more than 80 *kimarite*, or techniques, used to win a match. Taiho's *yorikiri* move is very popular because it is simply a technique whereby one *rikishi* drives his opponent out of the ring. A *rikishi* often grabs his opponent's *mawashi* (garment worn around the waist, like a thick belt or loincloth) during a *yorikiri*. Taiho was quite good at it; he used the *yorikiri* move in about 30% of his victories. The *yorikiri* is all about linear momentum, a topic we'll discuss shortly. Now, I want to tell you what *rikishi* eat to achieve their phenomenal size, and how that food gets turned into the energy needed for their bouts.

Pass Me the Chanko-nabe, Big-Bang Style!

Rikishi live and train in a *heya*, which is translated commonly as a "stable," even though *heya* means "room" in Japanese. Top *rikishi* found and run the stables. A rigid hierarchy requires that junior *rikishi* perform menial tasks like cooking and cleaning.[9] If you have ever seen a sumo competitor, you know that eating is a big part of sumo life. A staple of *rikishi* diet is *chanko-nabe*, which is a stew high in protein and carbohydrates. Junior *rikishi* prepare *chanko-nabe* with chicken, beef, fish, tofu, vegetables (mushrooms, burdock, carrots, etc.), soy sauce, potatoes, and other culinary delights. Top it all off with a bunch of rice and generous amounts of beer and you're good to go!

How does the body use all that food? From a physics point of view, we need to think about energy conservation. When the Big Bang[10] occurred, the universe had a certain amount of energy. Since we can define any reference point for potential energy that we like, call the total energy zero. We

---

[9] Scandals in recent years have rocked sumo in Japan. One much-publicized case involved the death of a junior *rikishi* in June 2007, most likely the result of hazing. See "Sumo stable boss axed for death," *The Japan Times*, Saturday, October 6, 2007. Scandals coupled with an infusion of foreign participants have hurt the popularity of a sport so strongly rooted in Japanese tradition.

[10] I am not referring to the collision of two *rikishi*!

can imagine all kinds of particles whizzing around with positive kinetic energies and feeling, for example, gravitational attractions with associated negative potential energies. Particles and their associated antiparticles can be created out of the energy in the vacuum; they can also destroy each other when they collide, releasing energy back into the vacuum. Gravitational attraction brings particles close together, forming galactic dust, planets, stars, and all the other cool stuff in the universe. We are lucky enough to be sitting on a ball of matter that formed at just the right distance from a star for conditions on that ball to be what they needed to be for life to form. In our sun, protons collide with other protons in fusion reactions that convert hydrogen to helium. Collisions are rare because protons, which all have the same sign of charge, don't exactly like being near one another. Thanks to some thermal energy and a little quantum-mechanical tunneling,[11] protons in the sun react, setting off a chain of reactions that lead to photons streaming out of the sun. It takes about a billion years to go from a couple of protons colliding in the sun's core to photons routinely hitting us on the back and giving us sunburn. That's fortuitous, since faster reactions would mean a shorter lifetime for the sun, meaning a quicker doom for us. The Earth is about four-and-a-half billion years old; the sun is a tad older. For each passing second, approximately four billion kilograms of matter get converted to energy in the core of the sun. We understand that conversion because Einstein—through his famous equation $E_0 = mc^2$, where E_0 is the rest energy of an object with mass m, and c is the speed of light in vacuum—showed us that mass is but a form of energy. When the sun turns into a red giant in five-and-a-half billion years or so, the Earth will either get swallowed up or have all its water and atmosphere boiled away. We still have plenty of time to figure out how we can save humanity!

What does all that stuff have to do with sumo? Those photons streaming at us from the sun are responsible for photosynthesis in plants, as well as all life on Earth. What physics is able to describe is the continual conversion of one form of energy into another, while always maintaining

---

[11] The sun is not hot enough for classical physics to explain how the protons get close enough for fusion. Quantum mechanics helps us understand how all this works, because some processes that are forbidden in classical physics actually have a small probability of occurrence. That's lucky for us!

the same total amount. By converting energies to other forms, order can be created out of disorder. I mentioned the second law of thermodynamics back in chapter 6. The local creation of order requires the global increase of disorder. One way we enjoy order on Earth is that atoms form molecules, which then form macromolecules like DNA, which then help form us. I'm obviously leaving out a lot about how we are formed, but you get the idea. Order gives way to disorder, as when things die. We enjoy forms of order on Earth, but that is partly because the sun will eventually die. Converting one form of energy into another while increasing order means that something has to do work, which in turn leads to some fraction of waste energy.

The idea here is that when a *rikishi* sits down to a nice bowl of *chanko-nabe*, he is about to eat something whose very existence (and that of the *rikishi* as well) is owed to the sun, and ultimately to the Big Bang. Thinking about all that is why it's a good idea to enjoy one's *chanko-nabe* with a cold beer! Where I want to take you next is a look at how energy conversions in our body allow us to do everything we do.

A Little Thermodynamics

I'll now quantify the first two laws of thermodynamics. The first law, which is nothing more than energy conservation, is

$$\Delta E = W + Q, \qquad (9.1)$$

where ΔE is the change in a system's internal energy (i.e. the total energy inside a given system) during a given process, W is the work done on the system during that process, and Q is the amount of heat flow during the process. Note that $Q > 0$ if heat enters a system and $Q < 0$ if heat leaves. I do not write Δ in front of W and Q because neither exists at a given instant in time. They have meaning only when we are discussing amounts transferred during a given process. It is meaningful, however, to think of a certain amount of energy existing in a system at a given time. Equation (9.1) tells us that if Q were 0, one could convert 100% of all input energy into work.

Before you think that we can build a car that converts all of the stored chemical energy in its gasoline into work, we need to write the second law of thermodynamics. I told you that closed systems tend to become more disorderly as time ticks away. I introduced briefly the physical quantity that measures disorder back in chapter 6. It is *entropy*, and it is denoted by the symbol S. The second law of thermodynamics tells us that in a closed system, entropy tends to increase over time. Expressed as an equation, the second law is

$$\Delta S \geq 0, \tag{9.2}$$

where the equal sign holds only for certain contrived processes. For most everything we see, and certainly the processes in the sports world, entropy increases. While simple in appearance, equation (9.2) puts constraints on every process that takes place in the universe. Heat flows when entropy increases, meaning there is always waste heat when we try to do work. Physicists often say that the first law of thermodynamics tells us that we can't win (more work out than energy in), while the second law tell us that we can't even break even (same work out as energy in). The requirement for some waste heat is why we don't have perpetual motion machines, or engines that dissipate no heat.

To think about how entropy works, consider a sumo stable. Suppose that the junior *rikishi* spend several hours cleaning it up. The stable by itself is thus more orderly and has less entropy than it did before. But the *rikishi* had their collective entropy increase more than the stable's entropy went down. After all, they expended energy and got all sweaty while working. The total system (of *rikishi* + stable) had its entropy go up. What happens to the stable over time as all the *rikishi* use it for meals, sleeping, bathing, cooking, training, and so forth? The stable gets more cluttered over time, no? That indicates to us that over time, if no *rikishi* clean it, the stable's entropy goes up.

The aforementioned idea applies equally to ordered molecules in cells. Left alone, a cell will become more disordered and die. If all the constituent atoms were to peel off and move in different directions, the "cell" would become very disorganized. To keep the cell living, chemical reactions must take place. Those reactions organize some atoms into molecules; the reactions also dissipate some heat energy. There is, of course, an overall increase in entropy as the environment surrounding the cell becomes more disordered than the cell is ordered.

Follow the Energy!

I'll now describe how energy goes from sunlight to the explosive burst of a *rikishi* at the start of a match. I'll outline the trail of energy because I don't want to turn this book into a biology text. I do, however, urge you to read more about the details behind what I am about to describe.[12] Biology and chemistry—both of them fascinating fields of study—tell us a lot about how our bodies work. Underlying those fields, as pretty much everything else, are the laws of physics.

Start with sunlight hitting the scrumptious veggies used in *chanko-nabe*. Photosynthesis is the process by which electromagnetic energy from the sun combines with water taken from the soil and carbon dioxide taken from the air to produce sugars. Waste products from photosynthesis are oxygen gas, which we use when we breathe, and a little heat energy. The presence of the heat energy ensures energy conservation (first law of thermodynamics) and demonstrates that the conversion of the sun's electromagnetic energy into stored chemical energy is not 100% efficient (second law of thermodynamics). The sugars produced contain stored chemical energy, and it's the energy stored in the chemical bonds of the sugars that a *rikishi* uses after swallowing a big bite of burdock root from his *chanko-nabe*.

The aforementioned process may be stated succinctly in a chemical equation as

$$(\text{electromagnetic energy}) + H_2O + CO_2$$
$$\rightarrow \text{SUGARS} + O_2 + E_{heat}. \tag{9.3}$$

Useful energy is stored in the sugars while an amount of energy, E_{heat}, becomes unavailable. To recover the stored energy, the arrow in the above equation must be reversed. The reversed process is called respiration. You

---

[12] A good cell biology book is *Essential Cell Biology* by Bruce Alberts et al. (Garland Publishing, Inc., New York, 1998).

can see why we need oxygen when we inhale and why we release carbon dioxide when we exhale. The release of energy from the food we eat through the reverse process in equation (9.3) is called "oxidation," because of the need to combine the sugars with oxygen.

In a qualitative sense, equation (9.3) and the reversed process of respiration are easy to write down. Those reactions actually occur via several reactions in which specialized proteins called enzymes act as catalysts. When a *rikishi* finishes off a bowl of *chanko-nabe*, he has added a bunch of sugars to his body from the food—say, from the carbohydrates in the accompanying rice. The *rikishi* then sits back and relaxes. All the sugars he has ingested, as well as the sugars already in his body, do not then undergo simultaneous respiration reactions in which all the sugars release their stored chemical energy, followed by a release of water vapor and carbon dioxide. That is, there has not been a wave of spontaneous combustion in the sumo stables! Respiration in our bodies takes place much more slowly, and in a more controlled way. Even though sugars contain stored energy, they do not release it on a whim. Sugars in fact would almost never release their energy without a little help from enzymes. Mother nature has provided us with numerous types of enzymes, each of which has a particular chemical reaction it helps commence. The cells in our bodies, wonderfully elaborate, contain a symbiotic group of enzymes that aid reactions when needed. A *rikishi* certainly does not need to have all of his stored energy released in one instant!

Once a sugar releases its energy, it is then stored for an incredibly brief time before a cell needs it. The molecules that store energy in their chemical bonds are called "activated carriers," the most important of which, at least for the discussion here, is ATP. The acronym stands for adenosine 5'-triphosphate, though we don't need to know what all that means to understand what follows. Energy conversion takes place in our cells' mitochondria, small bodies that reside in the cytoplasm between the cell nucleus and the cell's plasma membrane.

The most important sugar used by our bodies for energy release is glucose, which is represented as $C_6H_{12}O_6$. The respiration reaction looks like

$$C_6H_{12}O_6 + 6O_2 \rightarrow 6H_2O + 6CO_2 + (686\,\text{kcal of energy}). \qquad (9.4)$$

Photosynthesis works in the opposite direction. The 686 kcal (or 686 food calories; see footnote 10 in chapter 4) emerge when 1 mole of glucose is oxidized. A mole of glucose is roughly 6×10^{23} (that's six hundred billion trillion if you're scoring at home) glucose molecules, which has a mass of about 180 grams.[13] The released 686 kcal of energy is in the form of heat if the breakdown of glucose takes place in air.

In a cell, the reaction described by equation (9.4) allows for the biosynthesis of ATP from ADP, which is adenosine 5'-diphosphate (as with ATP, the molecular details of ADP are not needed here).[14] When the cell needs some energy, ATP is ready to release its stored energy. It does so through hydrolysis, which is a chemical reaction that splits water molecules into pieces that are useful for other reactions, with an ADP popping out. That ADP is then available for another reaction, to create more ATP. The cycle continues on and on in our living cells. Note that ATP holds its energy for only the briefest of moments. Once it has stored energy, ATP is ready to release it. We don't have a bunch of ATP hopping around in our cells waiting for long periods of time for something to come along and pilfer their stored energy.

One vital ingredient in equation (9.4) is oxygen. We of course obtain that via breathing. We see in the glucose reaction that 6 moles of oxygen are needed to gain 686 kcal of energy. A good approximation is to assume that oxygen at standard temperature and pressure is an ideal gas, i.e. a collection of non-interacting gas molecules. Chemistry tells us that a mole of ideal gas has a volume of about 22.4 liters. For each liter of oxygen we breathe, we can figure out how much energy is available to us by the following:

$$\frac{686 \, \text{kcal}}{(6 \, \text{moles of O}_2) \, (22.4 \, \text{liters/moles of O}_2)} \simeq 5 \, \text{kcal/liter.} \qquad (9.5)$$

For each liter of oxygen our body consumes, it uses about 5 food calories of energy. Note that our lungs do not retain all the oxygen we breathe in with each breath. About 16% of exhaled air is oxygen. That means

---

[13] To obtain the mass, simply add up the number of nucleons (protons and neutrons). C has 12 nucleons, H has 1 nucleon, and O has 16 nucleons. The mass of a mole of glucose is thus $(6 \times 12 + 12 \times 1 + 6 \times 16)$ grams = 180 grams. This calculation is based on the definition that 1 mole of carbon has a mass of exactly 12 grams.

[14] The glucose, oxygen, and ADP also require a phosphate group to create ATP. We need not get too deep into the biochemistry to gain a qualitative feeling for energy transfers in the body.

that our lungs, which keep none of the inhaled nitrogen (about 78% of air) keep only about 4–5% of the matter inhaled in a given breath. Depending on what our daily activities are, we need anywhere from 300 liters to 800 liters of oxygen each day. That amounts to 1,500–4,000 food calories burned per day, as calculated from equation (9.5).

Let's make sure now, that we know where we are. A *rikishi* gulps down a big bite of *chanko-nabe*. The glucose in the food uses a pinch of oxygen from the lungs and a morsel of ADP to create some ATP. The mitochondria play the role of the Iron Chef.[15] There are some waste products along the way, like the carbon dioxide that we exhale and the water that we dispose of in other ways. But again, the ATP stores energy. We are now ready to see one way in which our bodies use that stored energy.

I *Am* Working!

"Work," like energy, momentum, and power, is one of those words we use routinely. We "go to work," "work hard," order a big submarine sandwich with the "works," and then "work off" the tire around our middles. In physics, however, "work" has a very precise definition. It is the product of the component of force along the direction of motion with the displacement. "Force times distance" is a popular way to think of it, but be careful. Some forces, like the centripetal force we met in chapter 8, can alter an object's direction of motion, but still do no work. A centripetal force always points perpendicular to the displacement, meaning there is no component of the centripetal force along the displacement. The idea here is that in physics, work is performed only if there is some component of force in the direction of the displacement.[16]

To get a better handle on the physicist's idea of work, let's imagine a time when the great Taiho was training for a *honbasho*. One of the training

---

[15] The *Iron Chef* is my favorite Japanese television show. It pits guest chefs against "Iron Chefs" in battles to see who can create the best dishes in just an hour of cooking. The show ran mostly between 1993 and 1999.

[16] Calculating work can sometimes be a challenge. The formal definition involves a line integral of the component of the force in the direction of the displacement multiplied by a tiny displacement over a specified path. In some cases, changing the path changes the work performed. We will not need to worry about those details here.

devices he used, called a *teppo*, is somewhat like a telephone pole, but not so tall. Like a boxer's heavy bag, or the meat carcasses Rocky Balboa used in *Rocky*, the *teppo* helps condition *rikishi* for their bouts. As a *rikishi* hits the *teppo* with his hands, he strengthens his arms, legs, and back. He develops some pretty tough hands, too! Imagine Taiho hitting the *teppo* and leaving his hands on it. His legs are behind his center and he exerts a great deal of strength pushing on the *teppo*. The *teppo* does not move because it is entrenched firmly in the ground. Taiho does not move because the force from the *teppo* on him exactly balances the static friction force on Taiho from the ground. A junior *rikishi* observing off to the side would see no movement. If you want to try this yourself and you don't happen to be near a *teppo*, stand up and push on a solid wall. From the physics point of view, Taiho does no work on the *teppo*, and you do no work on the wall. "But wait," you exclaim, "I'm getting tired! How can I not be doing any work?" Neither Taiho's *teppo* nor your wall displaces, so no work is being done. You and Taiho could push on your respective inanimate objects for an hour and still do no work, even though you may both perspire and become very tired. Though it *is* true that no work is done on the *teppo* or on the wall, there *is* actually work being done. We will understand this only if we leave the macroscopic world of a *teppo* and a wall.

Our lungs help initiate a reaction by which we pick up oxygen. Our muscles reverse the reaction and release oxygen. We have seen how our bodies use glucose and oxygen to store energy in ATP. We now examine how our muscles use ATP and perform work on a microscopic level, even if we are pushing a wall on which we do no work. It's easier to understand how we do work on a ball we lift off the floor. We exert a constant force on the ball over a certain displacement and, if the force vector points in the same direction as the displacement, the work is simply the product of the force and the displacement. Inside our bodies, skeletal muscles contract and move our bones around. From an anatomical point of view, we can imagine muscles pulling on bones and displacing those bones, thus doing work. When pushing on a *teppo* or wall, however, we cannot see what is being displaced.

After we inhale, our lungs take oxygen from the air and transform it mostly to hemoglobin-saturated blood. Some of the blood makes its way to our muscle tissues. A muscle cell, or fiber, is an elongated structure with several nuclei. Although one muscle cell may be just 50 μm in diameter,

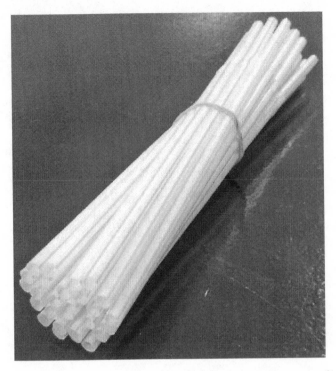

Figure 9.2. Toy model of a muscle cell. Each straw represents a myofibril within the muscle cell. The toy is not to scale, meaning I that do not have nearly enough straws bundled together (I would need several hundred!).

a muscle cell may be a good fraction of a meter long. Within the fiber are many *myofibrils*, just 1–2 μm in diameter, meaning that there may be from several hundred or so to a couple of thousand myofibrils within one fiber. Imagine a bunch of straws tied together to make a cylinder (see Figure 9.2). Within each myofibril (or straw) are things called *sarcomeres*. Think of taking a straw and drawing a bunch of roughly evenly spaced circles around it, with the plane of each circle being perpendicular to the longitudinal axis of the straw—almost like the spaces between the coins in a roll of coins (see Figure 9.3, and imagine each white segment as a coin). Each of those segments of the straw, or coin if you like, is a sarcomere. A sarcomere is only about 2.5 μm long (that would be the thickness of a coin in our imaginary roll of coins). Within each sarcomere are two types of filaments, the two capable of sliding along one another. One type of filament, which is thin, is the *actin* filament; the other, which is thick, is

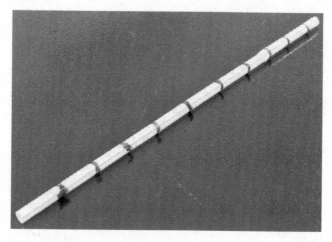

Figure 9.3. Within a given myofibril (straw) are many sarcomeres. I have sectioned off the sarcomeres (the white sections) with a black marker.

Figure 9.4. Within a given sarcomere there are thin filaments (actin) and thick filaments (myosin). My toy model uses a straw cut in half with vermicelli (whole wheat for color distinction) for the actin and spaghetti for the myosin.

the *myosin* filament. Both types of filaments run along the cylindrical axis of the myofibril. Try to picture (see Figure 9.4) a section of straw (a sarcomere) with some thick spaghetti (myosin filament) and some vermicelli (actin filament). A myosin filament is about 16 nm (1 nm = 10^{-9} m) in diameter, which is about three times wider than an actin filament.[17]

My guess is that you are thinking, "There is too much biology here! My brain is turning into spaghetti!" Stay with me just a little longer, because the punch line is right around the corner. Along the myosin filaments are *myosin heads*. See Figure 9.5; a myosin head (locked onto an actin filament here) looks like a caveman club. An ATP molecule hops onto a

---

[17] I took all of the distance measurements in this paragraph from the reference in footnote 12, above.

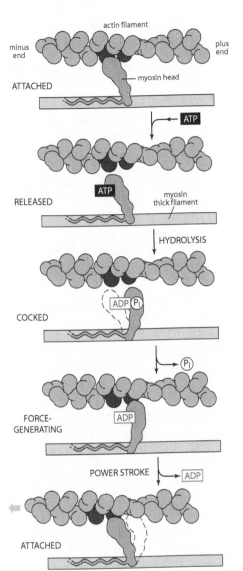

ATTACHED—At the start of the cycle shown in this figure, a myosin head lacking a bound nucleotide is locked tightly onto an actin filament in a *rigor* configuration (so named because it is responsible for *rigor mortis*, the rigidity of death). In an actively contracting muscle this state is very short-lived, being rapidly terminated by the binding of a molecule of ATP.

RELEASED—A molecule of ATP binds to the large cleft on the "back" of the head (that is, on the side farthest from the actin filament) and immediately causes a slight change in the conformation of the domains that make up the actin-binding site. This reduces the affinity of the head for actin and allows it to move along the filament. (The space drawn here between the head and actin emphasizes this change, although in reality the head probably remains very close to the actin.)

COCKED—The cleft closes like a clam shell around the ATP molecule, triggering a large shape change that causes the head to be displaced along the filament by a distance of about 5 nm. Hydrolysis of ATP occurs, but the ADP and P$_i$ produced remain tightly bound to the protein.

FORCE-GENERATING—The weak binding of the myosin head to a new site on the actin filament causes release of the inorganic phosphate produced by ATP hydrolysis, concomitantly with the tight binding of the head to actin. This release triggers the power stroke—the force-generating change in shape during which the head regains its original conformation. In the course of the power stroke, the head loses its bound ADP, thereby returning to the start of a new cycle.

ATTACHED—At the end of the cycle, the myosin head is again locked tightly to the actin filament in a rigor configuration. Note that the head has moved to a new position on the actin filament.

Figure 9.5. Work being done, on a small scale. (© 1998 from *Essential Cell Biology* by Bruce Alberts et al. Reproduced by permission of Garland Science / Taylor & Francis, LLC.)

myosin head. The ATP releases its stored potential energy by creating an ADP molecule and "cocking" the myosin head. The same idea is behind the bent ruler that I cocked on my kitchen table back in Figure 1.2. Part of the stored energy in the ATP molecule now resides in the stored energy of the cocked myosin head. We have reached the *coup de grace* because the myosin head is capable of doing work now! The cocked myosin head attaches to the actin filament and uncocks itself. Put another way, the myosin head exerts a force on the actin filament and moves it—the caveman club *does work* on the spaghetti, and the muscle contracts. We have found the work! We have gone from the Big Bang to fusion in the sun, to photosynthesis, to *chanko-nabe*, to the intake of oxygen, to the oxidation of glucose, to storing energy in ATP, to releasing that energy into a cocked myosin head, to the myosin head uncocking and doing work on an actin filament. Whew!

By the way, when you get tired of pushing on a wall, you might perspire. You do so to help release heat from your body. That heat is a waste product of the aforementioned energy conversions in your muscles. Skeletal muscles are only about 20% efficient (±6% or so, depending on the activity).

For a *rikishi* to maintain a roughly constant tension in his muscles while he pushes on a *teppo*, myosin heads are moving actin filaments several hundred times per second. The actin filaments slide back after the myosin heads have moved them. When these filaments slide back, they uncoil (ever so slightly!) like a spring, thus releasing a little bit of stored potential energy. This idea is behind the antagonistic muscles in the body, the muscles that perform in opposition to the movements of agonist muscles. The myosin heads act only to contract a muscle, i.e. tighten it; they cannot relax a muscle. Imagine doing an arm curl. Your biceps (agonist) contract, meaning that the caveman clubs (myosin heads) are in action. At the same time, your triceps (antagonist) stretch; the caveman clubs are doing nothing in the triceps. When you drop the weight by uncurling your arm, your triceps (agonist) contract while your biceps (antagonist) extend. Without antagonist muscles, we could not undo what an agonist muscle does when it contracts.

I have had to leave out many biochemical details, but I hope you now have an appreciation for the fact that we cannot escape the laws of physics, whatever we happen to be doing. Even if a *teppo* (or wall) does not move when pushed, there must be work being performed somewhere. We zoomed in far enough to get to Figure 9.5, where we discovered a force displacing an object, thus doing work. I could zoom in still farther and describe the structure of the ATP molecule and the myosin head. And after moving past the molecular level, I would be telling you what the electrons on individual atoms are doing. The force of the myosin head on the actin filament is ultimately an electromagnetic force.

There is obviously much more to this story, and I urge you to read more about the fascinating, often amazing way in which our bodies work. For now, I'll return to some more physics, once again on a macroscopic scale.

Linear Momentum

Now that we understand how energy gets transformed into work in the body, we are ready for the *tachi-ai*, which is the initial charge of the *rikishi*. Once Taiho and Asashio reach an unspoken agreement during the *niramiai*, they charge at one another. Refer again to Figure 9.1 and note that there is not much space separating Taiho and Asashio when they begin their charge. What that means is that neither combatant will be able to generate much speed by the time of the collision. This is why it helps to gobble up one's *chanko-nabe* and put on some serious weight.

To understand why a beefy body is important, I need to tell you about linear momentum. We met angular momentum back in chapter 5, in the context of rotational motion. We now focus on translational motion, i.e. motion along a straight line. Newton's second law is often stated as I gave it in equation (3.6),

$$m\vec{a} = \vec{F}^{\,\text{net}}.\tag{9.6}$$

When Newton first laid down his famous second law, he thought not of acceleration but of linear momentum. The formal definition of linear momentum, given by the symbol $\vec{p}$, is

$$\vec{p} = m\vec{v}.\tag{9.7}$$

Looks simple enough, doesn't it? There is actually more in that equation than meets the eye, as you might imagine. You have probably heard "momentum" used in common vernacular, such as, "Her momentum carried her forward." Though that statement is not correct from a physics point of view, it does bear some relation to the laws of physics. Think about the product of mass and velocity given in equation (9.7). Would you rather be hit by a ping-pong ball (mass 2.7 grams) smashed at 100 km/hr (~62 mph) or a baseball (mass 145 grams) pitched at that same speed? Both probably hurt, but the ping-pong ball is my choice, given that it is nearly 54 times less massive. By equation (9.7), then, the ping-pong ball has nearly 54 times less linear momentum.

To see why the ping-pong ball is the better choice, I need to rewrite Newton's second law in the form Newton originally wrote it (with cleaner and more modern notion, of course). A more fundamental way to write Newton's second law is

$$\frac{\Delta \vec{p}}{\Delta t} = \vec{F}_{ave},\qquad(9.8)$$

where $\vec{F}_{ave}$ is the average net external force acting on a particle during the time interval Δt.[18] When you note that average linear acceleration is

$$\vec{a}_{ave} = \frac{\Delta \vec{v}}{\Delta t},\qquad(9.9)$$

you can see that equation (9.8) emerges when you combine equations (9.7) and (9.9) with Newton's second law as it is written in equation (9.6). What equation (9.8) tells you is that the time rate of change of an object's linear momentum is the average net external force on it during the time that that net external force interacts with the object.

Let's go back to Taiho charging. If you survived his stare during the *niramiai*, you have but a second to decide what to do during the *tachi-ai*. To stop Taiho's charge, you must reduce his speed to zero, meaning you also need to reduce his linear momentum to zero. Rearrange equation (9.8) and get

$$\Delta \vec{p} = \vec{F}_{ave} \, \Delta t.\qquad(9.10)$$

Physicists refer to either side of the above equation as the "impulse" of an interaction, yet another term you have probably used before in common

---

[18] If you have studied calculus, you know that the instantaneous version of equation (9.8) is $d\vec{p}/dt = \vec{F}^{net}$.

vernacular. The force on an object is often very complicated during a collision. That's why I use the definition of impulse with the average force.[19] If you want to stop Taiho in his tracks, you need the magnitude of Taiho's $\Delta \vec{p}$ to be the product of his mass and the speed he has when he first makes contact with you. This is because Taiho's final linear momentum is zero if you are able to bring him momentarily to rest. If you want to stop Taiho, you cannot control the left-hand side of equation (9.10). Ah, but you *can* control what goes on with the right-hand side of that equation. Force times collision time is a fixed product, but you are free to choose the force and the collision time you want, by way of attaining that fixed product. Thus if you don't think you can exert great force, then you need to extend the collision time. After Taiho makes contact with you, start backpedaling just a little. But don't go too far or you'll be out of the ring. If you feel that you can't move much farther back, you are going to have to exert a large force on Taiho to stop him. Now you see why *rikishi* need to be strong, not just large.

By the way, you have most likely made use of equation (9.10) even if you didn't realize it at the time. Have you ever jumped off a chair or out of a tree? You instinctively bent your knees when you hit the ground, didn't you? If you keep your knees straight, the time needed to bring your linear momentum to zero is so short that the force required is great enough to damage your knees. You bend your knees to extend the collision time, thus reducing the force needed to bring you to rest. A baseball glove works on the same principle. The padding serves to extend the collision time. And if you have ever caught a baseball without a glove, you most likely let the ball hit your hand while simultaneously pulling your hand back. That's a lot better than catching a ball bare-handed with no hand movement at all.[20] Perhaps you have had your life saved by an air bag. Physics rescued you! The air bag extends the collision time, reducing the force needed to stop you.

---

[19] Again, if you are comfortable with calculus, the right-hand side of equation (9.10) must be replaced with an integral of the time-dependent force over the time of the interaction.

[20] My wife found this out first hand, so to speak, at a baseball game in Arlington, Texas, back on Saturday, July 8, 2000. Ryan Klesko of the San Diego Padres lined a scorcher into the seats in foul territory adjacent to first base and nailed my wife's right hand.

The beautiful way in which the world works includes the idea that linear momentum is conserved for systems in which the net external force is zero. You can see that equation (9.10) requires that that be so if the force on the right-hand side is zero. The subtlety here is the meaning of the word "system." If Taiho is the sole entity in the system, then his linear momentum is certainly not conserved, since he feels a net force during a collision. If, however, the system includes both Taiho and Asashio, we are close to a situation in which linear momentum is conserved. Sure, the *rikishi* feel outside forces from the Earth (weights); but those forces are mostly canceled by normal forces from the ground. Each *rikishi* feels a frictional force from the ground; but when considering the system to consist of both *rikishi*, the frictional forces are mostly in opposite directions, and the magnitudes are comparable. Besides, the collision time when the *rikishi* meet is so short (well under a second) that the force one exerts on the other is the dominant force each *rikishi* experiences. But Newton's third law says that the force the first *rikishi* exerts on the second is equal and opposite to the force the second exerts on the first. Canceling those forces means that the net force on the system of two *rikishi* is essentially zero during the brief collision time. Newton's third law and the law of conservation of linear momentum are inextricably linked.

To analyze a collision in a more quantitative way, consider what change in linear momentum for a system really means:

$$\Delta \vec{p}_{\text{system}} = \Delta \vec{p}_{\text{Taiho}} + \Delta \vec{p}_{\text{Asashio}}, \tag{9.11}$$

or

$$\Delta \vec{p}_{\text{system}} = \left(\vec{p}_{\text{Taiho,final}} - \vec{p}_{\text{Taiho,initial}} \right)$$
$$+ \left(\vec{p}_{\text{Asashio,final}} - \vec{p}_{\text{Asashio,initial}} \right), \tag{9.12}$$

where I have stated explicitly linear momenta at the split second before a collision (initial) and the split second after a collision (final). If the system's linear momentum does not change, the above equation may be written as

$$\vec{p}_{\text{Taiho,initial}} + \vec{p}_{\text{Asashio,initial}} = \vec{p}_{\text{Taiho,final}} + \vec{p}_{\text{Asashio,final}}, \tag{9.13}$$

i.e. the total system linear momentum before the collision is equal to the total system linear momentum after the collision. Remember that all those p's are *vectors*. Direction is just as important as magnitude. Suppose Taiho charges east and Asashio charges west; suppose further that the magnitude of Taiho's initial linear momentum is greater than Asashio's. That means that the system's linear momentum points east—and the system's linear momentum must also point east just after the collision. If the two combatants stick together after the collision, just think of the two *rikishi* as a single conglomerate object, a big one. Once Taiho spun Asashio around in their famous 1960 match, linear momentum was on Taiho's side. Asashio's center of mass was essentially above his feet and he could not drive Taiho. Unable to drive much, Asashio had no choice but to let Taiho set the system's linear momentum direction. Both men left the ring; Taiho won, because Asashio stepped out first.

You now understand why the *niramiai* and *tachi-ai* are so important. If your nerves waver under the stare of your opponent and you lose the linear momentum battle at the initial collision, you may find yourself heading backwards and out of the ring. Or, like Asashio, you could be spun around and driven out in the direction that was originally behind you. You could lose in a matter of seconds! Since there is not much space in which to attain a great initial speed, you can see why being BIG helps. Equation (9.7) tells you that if you want a large linear momentum, and you can't attain a great speed, you need a large mass. Please don't think, however, that *rikishi* are just obese men. Top *rikishi* may have big guts, but there is a lot of muscle in there. If you think the great size of a *rikishi* impedes his flexibility, watch him do a *matawari* split. That exercise, in which a *rikishi* sits on his derriere and moves his legs as far apart as he can, may astound you. I don't feel so flexible when I watch a *rikishi* perform that exercise!

I do not have the space in this book to look at the other 80 or so *kimarite*. I urge you, though, to look at books on sumo (like the one I mentioned in footnote 4) and/or on the web for complete descriptions of all the various moves. What you have seen up to now in this book is more than enough to understand the physics behind the various *kimarite*. All you need is a basic understanding of Newton's second law for translations, the corresponding law for rotations, and a good grasp of linear and angular momentum.

We now say *sayonara* to the world of sumo. It's time for the post-game show, where I'll offer you something that may make watching sports a little more exciting.

10

The Post-Game Show

My primary hope for readers of this book who have come this far is that they find themselves having a little more fun watching and playing sports, because they have a little bit better understanding of how the world works. I certainly haven't plumbed every sport, much less every facet of the few sports I did hit, and there is also a lot of physics that I have not included in this book. But you can still apply the major conservation laws I have discussed (energy, linear momentum, and angular momentum) to what you see going on in the world. And of course you need not limit your scope to sports. A knowledge of energy conservation and the laws of thermodynamics, for example, will help you to understand better many of the issues surrounding current political debates on energy (at least from a scientific point of view—I can't help you with the political point of view). The beauty of the universe is that the laws of physics employed by the great scientists picking up their gold medals in Stockholm each December are the same laws that govern the motions of all of us. The angular momentum conservation law used by high-energy physicists to analyze a complex subatomic process is the same law that explains for us why Katarina Witt speeds up when she pulls her arms in close during a spin. Amazing, huh?

I will offer you one more tidbit in this final chapter. There are no applications of physics here. What I will be showing you is a way you can use modeling to add a little enjoyment to watching sports. You've noticed from what you've seen in this book that physicists make extensive use of models. I created computer models using the laws of physics for many of the topics I covered. Revealing a graph of the trajectory of a discus, for example, is nothing but my showing you a computer model of a physical phenomenon. What I will show you now is a computer model I use for a college football prediction contest. Like a few others in my family, I love college football, and our weekly viewing in the fall is enhanced by participation in our prediction contest. If you are a fantasy player, you know what I mean. Fantasy sports are, for the most part, just computer models.

If you and a friend are about to watch a college football game (or any other sport for that matter), what might you argue about before the game starts? You will discuss, or perhaps debate with some vigor, who will win the game. There are newspaper contests in which you can check off the teams you think will win a given set of games. There are also venues where you can place legal bets on games.[1] Some contests use points scored in a given game as either tie-breakers or the basis of the contest itself. What I set out to create about a decade ago was a way in which I could include both the "Who will win?" part of a game with the "How many points will a given team score?" aspect.

The *S* Equation

I'll give you the heart of our contest first, and then I'll try to elucidate the various pieces. The symbol S represents one's "score" for a given game. The simple equation for S is

$$S = P \cdot R - 1, \tag{10.1}$$

where P is the "prediction" factor and R is the "rankings" factor.

Let's start with P. Suppose team A plays team B. To compete in our contest, you need to predict the score of the game, meaning that you need to tell me how many points you think team A will score and how many points you think team B will score. There are endless ways one could quantify the quality of one's score prediction. The one I chose, after some deliberation, makes use of the percent difference of one's prediction from the actual score. Percent difference is simply the absolute value of the difference of two numbers divided by their average and multiplied by 100%. Suppose you pick P_A points for team A and P_B points for team B. Suppose

---

[1] My family and I do not incorporate money into our college football prediction contest in any way. It's purely for fun. Having stated that, I am in no way so naïve not to realize that gambling is a multi-billion-dollar industry.

further that the actual score of the game is P'_A for team A and P'_B for team B. The percent difference between your pick for, say, team A and team A's actual score is

$$(\% \text{ difference})_A = \left| \frac{P'_A - P_A}{\frac{1}{2}\left(P'_A + P_A\right)} \right| \cdot 100\%. \qquad (10.2)$$

Then define the percent difference for team B in the same way. As an example, suppose you predict 10 points for team A and that team scores 15 points. The percent difference is 40%. Note that if you had predicted 20 points, the percent difference would have been about 29%. Although you miss the actual total by the same five points as before, your percent difference is lower. My idea was that missing low is worse than missing high, where "missing" can refer either to one's pick or to the actual score. By that I mean that equation (10.2) is symmetric under interchange of P_A and P'_A, meaning that you get the same 29% difference if you predict 15 points and team A scores 20 points.

The reason I think missing low is worse than missing high is because I think it's is bolder to predict a low score than it is to pick a high one. The ultimate example is a shutout. If I predict a shutout, I'm thinking that a team can go an entire game and not even get a lousy field goal. That's a pretty bold pick! If I pick zero points and get it right, equation (10.2) has a little trouble with a big zero in the denominator. In that case, I set the percent difference to zero. Note, too, that for any nonzero prediction you make, you get the same percent difference if a team gets shut out (or vice versa—you predict a zero and a team scores any number of points). In that case, equation (10.2) gives 200%. The shutout is the one special score in our contest. If you predict it and it comes true, you reap a nice reward. It's the one score where predicting 3 points is the same as predicting 40. When it comes to the number 0, all other numbers are the same multiplicative factor away! As scores move away from zero, percent differences go down. Be warned, though. If you think picking high will take you to victory, you still need to know on which side of the actual score to be. If you pick 30 points for a team and I pick 10 points, you do better if the team scores 20 (40% for you and 67% for me), but I do a lot better if the team scores 7 (124% off the mark for you and 35% for me).

My philosophy is that if a powerhouse plays a weak opponent, I don't think there is much fun in trying to predict the gobs of points scored by the powerhouse. If I predict 42 points for, say, Ohio State in one of its early-season matchups against a less-talented Ohio school, I have made a decent pick even if the Buckeyes score 49 points. (I predicted six touchdowns and they scored seven. Ho hum.) If, on the other hand, I predict 24 points for Ohio State in its annual clash against Michigan, and Ohio State actually scores 17 points, the seven points I miss in that game are bigger, it seems to me, than the seven I missed in the earlier blowout. The percent difference is 15% in the blowout and 34% in the Michigan game. The seven points I missed in the latter game simply represent a bigger chunk of Ohio State's actual score than the seven I missed in the former game.

To compute the P factor in equation (10.1), I thus need your predictions for how many points teams A and B will score (P_A and P_B). I also need to know the actual score (P_A' and P_B'). My prediction factor is then computed from

$$P = \left[1 + \frac{1}{2}\left(\left|\frac{P_A' - P_A}{P_A' + P_A}\right| + \left|\frac{P_B' - P_B}{P_B' + P_B}\right|\right)\right]^5. \tag{10.3}$$

You may be asking, "Where the heck did you get THAT?!" There is no science here! There is only a little art. I cannot *derive* the above equation. All I can do is tell you that I played around with several versions of P before settling on the one you see. The percent differences are lurking in there; I have, however, altered the factors. The prediction factor gives $P = 1$ for a perfect pick. The power of 5 in the exponent is again evidence of my tinkering with the equation.

Let's move on to the rankings factor, R, in equation (10.1). What I wanted to do with that factor was quantify upsets. Which is a bolder prediction: picking Appalachian State to beat Michigan (which happened on September 1, 2007) or picking Michigan to beat Ohio State (which did not happen on November 17, 2007)? The first game shocked the nation, but the second was not a big surprise. You are rewarded nicely in our contest if you are able to predict a monster upset; you are also penalized harshly if you *miss* a big upset (or predict a big upset and get it wrong). The P factor says nothing about winners—it is, after all, symmetric under interchange of A and B. If you correctly predict the winner, $R = 1$ and that's it. If, however, you are wrong, $R > 1$.

To quantify the magnitude of an upset, I need a rankings system. We use the Congrove computer rankings.[2] If you predict, correctly, #1 to beat #2, you get the same R you would if you were to pick correctly #1 to beat #40. But in cases where you are wrong, the penalty is much worse in the latter game than in the former. I do this because the rankings of two close teams are essentially interchangeable. There is really no difference between #1 and #2. There *is* a difference between being ranked #1 and being ranked #40. I want to reward the person who goes out on a limb and picks #40 to beat #1, and gets it right. At the same time, I want to penalize the person who is crazy enough to make the same pick, #40 to beat #1, and gets it wrong. Pick your upsets wisely!

To quantify the rankings factor, you need rankings for the two teams involved in a given game. Call R_A and R_B the rankings for teams A and B, respectively. I define R, the rankings factor, as

$$R = [1 + (R_A - R_B)^2]^{1/3}. \qquad (10.4)$$

As with the expression for P, there is nothing scientific about my choice for R. I wanted something that grows as the rankings difference between two teams grows. The rate and form of that growth is simply my personal choice. The minimum value of R is about 1.26 when teams differ in ranking by one; the maximum is about 24 if they differ in ranking by 120 (see footnote 2). Figure 10.1 shows a plot of R versus ranking difference. Note that R rises fast at the beginning and then becomes essentially linear. Keep in mind that R is computed from equation (10.4) only if one does not pick the correct winner. If you get the winner right, you get $R = 1$.

To compute your score, or S, for a given game, compute the prediction factor P from equation (10.3) and the rankings factor R from

[2] Visit the web site at www.collegefootballpoll.com/current_congrove_rankings .html. I actually prefer the Sagarin rankings (see www.kiva.net/~jsagarin/sports/ cfsend.htm), but we like to use a ranking system with just the 120 Division I-A (now called, I'm sad to say, the "NCAA Football Bowl Subdivision") schools. For the I-AA (now called the "Division I Football Championship Subdivision"—I'm not making this up) schools that appear in our contest, we give them all a ranking of 121. Some of the 125 or so I-AA schools are certainly better than some I-A schools, but we usually stick with top 25 games in our contest; for the most part, I-AA schools appear in our contest early in the season.

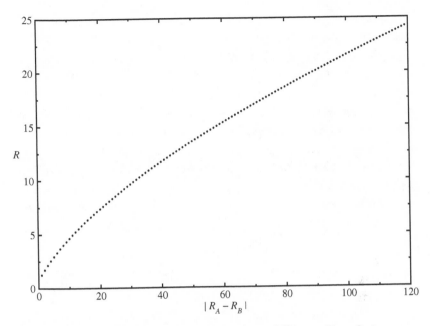

Figure 10.1. Rankings factor R versus rankings difference $|R_A - R_B|$.

equation (10.4). When you have P and R, use equation (10.1) to get S. If you make a perfect pick, meaning that you picked the exact scores for the two teams involved in a given game, you get $S = 0$. That's why I subtracted 1 from the product of P and R; perfection is a zero. The worse your score prediction gets, the larger S gets. The greater the rankings difference on a game whose winner you miss, the more you are hurt by a large S value. You must do two things well to score a low S value: pick the right winner and make a quality score prediction.

Despite the number of words I have used to describe the ingredients needed to compute S, it's all pretty easy to set up on a spreadsheet like *Excel*. The beauty of cut-and-paste is that once you program equations (10.1), (10.3), and (10.4) for one game, you never have to do it again. I have two spreadsheet cells for the rankings of the two teams playing each other. I also have two cells for my score prediction. I then use two more cells for the actual game score. Though it's not necessary to break up the S equation into its separate pieces, I do so in my spreadsheet. I have a cell for P, one for R, and another one for S. The six cells for rankings, score prediction, and actual game score require me simply to enter numbers. I must enter equation (10.3) for P in one cell and equation (10.4) in another cell. That's as tough as it gets, because the cell for S is trivial ($P \cdot R - 1$). If you don't mind taking a few minutes to get P and R set up, you can set this prediction contest up in no time. We pick typically 25 games a week. Everyone sends me their picks, I enter them into the computer, and then I enter the actual scores after the games are finished. My computer determines everyone's S value for each game, and I average the results to produce each participant's average S value per game. For added fun, we weight bowl games twice as highly as regular-season games. When the dust clears at season's end, the one with the lowest average S value is our contest champion.

Examples

Much of what appears in the previous section may seem a little abstract. Let's look at some real games and put some actual numbers in front of the abstraction. I'll use predictions for four people: Jay, Jarrett, Mandy, and me (Eric). I mentioned Appalachian State's big win over Michigan in the Big House on Saturday, September 1, 2007. Table 10.1 shows score predictions for that game as well as the actual result. (Note that since there were 119 I-AA teams in 2007—one more was added in 2008—Appy State's ranking was set to 120.) As you see from our predictions, we all thought Michigan would triumph, as did most everyone else. Appy State's shocking win gave us all a nice bath! Note that my pick was the worst, because I predicted the most points for Michigan and the fewest for Appy State. I earned the poorest S value with the poorest pick. Although

TABLE 10.1. Predictions and results for Appalachian State at Michigan (9/1/07)

	Mandy	Jay	Jarrett	Eric	Actual result
Appy State (120)	17	14	17	12	34
Michigan (21)	28	37	30	39	32
S value	52.25	62.90	48.54	74.99	

Note: Pre-game Congrove rankings are in parentheses after school names.

TABLE 10.2. Predictions and results for Texas at Iowa State (10/13/07)

	Mandy	Jay	Jarrett	Eric	Actual result
Texas (23)	31	38	37	34	56
Iowa State (113)	13	10	14	17	3
S value	5.55	3.74	4.89	5.92	

Note: Pre-game Congrove rankings are in parentheses after school names.

Division I-AA Appy State was not expected to win, they were certainly better than the 120th-ranked team in the country (see footnote 2). They came into the Big House as the two-time defending I-AA champions, and they captured their third straight title in December 2007. Upsets like this one are rare. Of course, had one of us picked Appy State, that person would have reaped a huge reward for such a gutsy pick. You can see from the size of S that Appy State's big win played a significant role in that week's contest.

Let's look at another game, one that is more typical of what one encounters. Table 10.2 shows a game for which we all picked the correct winner: Texas over Iowa State on Saturday, October 13, 2007. We all knew Texas would blow out Iowa State, but picking the weaker team's total was tough. Home field advantage did not mean much in that game! Note that Jay's pick was the best, because of his low-score prediction for Iowa State and his high-score prediction for Texas. Mandy edged me out because her getting closer to the lower point total of Iowa State was better than my getting closer to the higher point total for Texas. It's not that hard picking a lot of points for a monster favorite. It was a lot harder to predict that Iowa State could muster only a second-quarter field goal.

Okay, let's move on to a game where picking the winner was much tougher. Cincinnati played at South Florida on Saturday, November 3, 2007. Table 10.3 shows how we did. Cincinnati's road win was a mild upset, but only Jarrett managed to pick the Bearcats to win. He thus earned the best (lowest) S value. Note how much better Mandy's score picks were compared to Jay's. She was closer to both teams' actual scores and consequently earned a lower S value than Jay did.

I hope you're getting a feeling for the S equation. Earning a low S value not only means picking a game's winner, but having quality score predictions as well. This is my way of moving beyond the simple newspaper

TABLE 10.3. Predictions and results for Cincinnati at South Florida (11/3/07)

	Mandy	Jay	Jarrett	Eric	Actual result
Cincinnati (18)	24	16	29	20	38
South Florida (14)	28	21	22	25	33
S value	4.26	9.10	1.17	6.07	

Note: Pre-game Congrove rankings are in parentheses after school names.

TABLE 10.4. Predictions and results for Ohio State vs. LSU (1/7/08)

	Mandy	Jay	Jarrett	Eric	Actual result
Ohio State (1)	21	17	34	23	24
LSU (6)	24	24	21	27	38
S value	0.98	1.47	7.35	0.58	

Note: Pre-game Congrove rankings are in parentheses after school names.

contest and differentiating between people who all pick the same team to win a given game. Quality of score predictions provides an added layer of complexity—and an added layer of fun. It is even possible, though rare, to earn a better S value when picking the wrong team to win than someone else, who picks the right team, earns. If the rankings difference is small, the teams are virtually interchangeable; i.e. there is no real upset possibility. In that case, how well one can predict scores is the main ingredient in determining who gets the best S value. (If that bothers you, there's always the newspaper contest.)

I'll leave you with one last example, one that shows some small values of S. The final college football game played before I finished this book was the BCS Championship Game on Monday, January 7, 2008. LSU rolled over Ohio State for the national title. Table 10.4 shows how we did on that game. Jarrett, who went with the Buckeyes, had the worst pick. I hope you can see that my pick was the best. Sometimes you have a good feeling for a game, sometimes you don't. Seeing it all play out is a lot of fun.

To close this discussion, I want to remind you again that there is no science in the S equation I created. I do not claim that my equation is the best that one could use to model score predicting. You can quibble with my choice of an exponent here or a factor there. That's great. I enjoyed creating the S equation; I spent some time tweaking it until I thought it did a decent job of doing what I wanted it to do. The *S* equation does, however, contain my philosophy of what is important about predicting games. Over the past decade, Mandy, Jay, and Jarrett have all contributed to making our contest run smoothly. We have had our share of debates over what rankings system to use, how to count bowl games, and the like. You are welcome to take what I have offered here and use it for your own contest, and I encourage you to play with it and modify it as you see fit. Talk it over with your friends and see what ideas they can come up with. You won't offend me if you don't like my equation, and want to start from scratch! If you can produce something that makes watching sports a bit more entertaining, go for it.

A Final Word

Whether you are comfortable with it or not, you experience physics every day and every moment of your life. You are constrained by its laws and liberated by its contributions to technology. If men and women throughout humanity's history had not stopped to ponder how the universe works, you would not have air conditioning in your house, you would have no car to drive, you could not turn on a light at night, and you certainly could not watch sports on television. As a physicist, I am passionate about my chosen profession and biased about its contributions to our world. Always remember that science is ultimately about the pursuit of knowledge. Human beings apply scientific understanding toward technological advancements (and much else). Einstein told us that mass is a form of energy. Men and women decide how to apply that understanding to technology, whether nuclear energy to power a city or a nuclear bomb to level a city. An old maxim that I like is, "Just because we *can* does not mean that we *should*." Rather, embrace science as a way to understand the world around you. Your ability to think is your most potent asset.

Sports provide a wonderful respite from our high-tech or high-drudgery lives. Knowing something about why a ball moves the way it does can make that respite just a little more enjoyable. When you summon up that little kid inside of you, the one who makes sports so much fun, don't forget about the scientist within you, the one who was so vocal when you were that little kid.

FURTHER READING

Physics

Much of the physics in this book covers material often discussed in an introductory physics course, specifically the sections on mechanics.

Westfall, Richard. *Never at Rest: A Biography of Isaac Newton*. Cambridge: Cambridge University Press, 1981.

> Isaac Newton is one of the greatest scientists of all time. This is an accessible—and award-winning—biography.

Wolfson, Richard. *Essential University Physics*. 2 vols. San Francisco: Pearson/Addison-Wesley, 2007.

> A reasonably priced introductory physics book that is both succinct and well written.

Zimba, Jason. *Force and Motion: An Illustrated Guide to Newton's Laws*. Baltimore: Johns Hopkins University Press, 2009.

> A gentle introduction to Newton's laws of motion which focuses on building conceptual understanding through extensive use of illustrations.

Sports

There is a growing number of books devoted to the physics, or mathematics, of sports.

Gay, Tim. *The Physics of Football*. New York: HarperCollins Publishers, 2005.

> Football fans should consider this book, which delves deeply into the basic mechanics of America's sport.

Lipscombe, Trevor. *The Physics of Rugby*. Nottingham: Nottingham University Press, 2009.

> The rest of the world's equivalent of American football, explained using a level of physics similar to that in this book.

Fontanella, John J. *The Physics of Basketball*. Baltimore: Johns Hopkins University Press, 2006.

> The physics of basketball is discussed at a level similar to that in this book.

Adair, Robert K. *The Physics of Baseball*. 3d ed. New York: HarperCollins Publishers, 2002.

> This is the classic on baseball physics, and it is very readable.

Watts, Robert G., and A. Terry Bahill. *Keep Your Eye on the Ball*. New York: W. H. Freeman and Company, 2000.

> A slightly more technical, but just as enjoyable, book on baseball.

Haché, Alain. *The Physics of Hockey*. Baltimore: Johns Hopkins University Press, 2002.

> If you fancy the puck, check this one out.

Jorgensen, Theodore P. *The Physics of Golf*. 2d ed. New York: Springer-Verlag, 1999.

> A reasonable treatment of golf physics, coupled with some technical appendixes.

Armenti, Angelo, Jr. *The Physics of Sports*. New York: Springer-Verlag, 1992.

> A great collection of technical articles on sports physics, including many from the *American Journal of Physics*.

Wesson, John. *The Science of Soccer*. Bristol: Institute of Physics Publishing, 2002.

> This book covers most of the basics of soccer science at an introductory level, with more technical details presented in the final chapter.

Goldick, Howard D. *Mechanics, Heat, and the Human Body*. New Jersey: Prentice-Hall, Inc., 2001.

> An elementary textbook on the physics one needs to pursue a career in physical therapy.

De Mestre, Neville. *The Mathematics of Projectiles in Sport*. Cambridge: Cambridge University Press, 1991.

> A book at the level of differential equations which focuses on projectile motion in sports.

Barr, George. *Sports Science for Young People*. New York: Dover, 1990.

> If you are looking for a book to give a young person, try this elementary text.

Gardner, Robert. *Science Projects about the Physics of Sports*. Aldershot: Enslow Publishers, Inc., 2000.

> Another good one for young folks.

Suds

Denny, Mark. *Froth!: The Science of Beer*. Baltimore: Johns Hopkins University Press, 2009.

> Watching sports with a nice, cold pint of your favorite beer can be a sublime experience. My favorite is Guinness. Drink responsibly, and read Denny's book — it's packed with great information about beer.

INDEX

Page numbers in *italics* refer to illustrations and tables.

car (*continued*)
55; Newton's first law, while driving,
157, 157n8; rolling resistance and, 57;
streamlined, 21; testing drag force
with hand, 56, 164–65
Carlos, Roberto, 121, 144
cat, falling, 111
centripetal force, 158, 158n9; road
friction and, 159; work and, 185
chanko-nabe, 178, 180, 182–83, 185,
190–91
charge: conservation of, 75; dipoles and,
93–94, *94*
college football prediction contest, 197;
prediction factor, 199; rankings factor,
200–201, *201*; sample games, 202–4
compact disc, 109
corner kick, 123–24, *125*, 142; direct
goal from, 139, *140*; parameter space
for successful, *142*; rule of thumb for
success for, 142; schematic overhead
view of, *141*; trajectory of, *143*

density, air, 55–56, 56n6, 61, 113, 130,
132, 166; in Mexico City, 87
dipole: electric, 93; schematic of
interaction between two, *94*
Discobolus, 145–47, *146*
discus: launch angle for maximum range
of, *169*, 171; cross-sectional area of,
162, *163*; drag coefficient of, *167*;
forces on, *165*; lift coefficient of, *167*;
mass of, 151; optimum trajectory of,
173; range maximum as a function of
wind speed, *170*; technique for
holding, *154*
discus throw circle, *152*
displacement: angular, 78; calculus and,
36n12, 185n16; components of *25*,
26; spring, 12n15; vector, 24–25,
24; work and, 185–86
dohyo (sumo ring), 175–76
drafting, 63, 63n13
drag: air resistance and, 55; coefficient
of, 56; equation for, 55; water
resistance and, 113–14

echolocation, 4–5
Einstein, Albert, 54, 179, 205

Elway, John, 16, 19n7
energy: cars and, 55, 57; chemical, 101,
182; conservation of, 73, 75, 98–100,
106, 117–18, 154–55, 178–79, 190;
conversion between various forms of,
101, 179, 182–83, 190; elastic, 105–6;
mass and, 179; motional, 3; potential,
59; rotational kinetic, 100; sun, 179;
translational kinetic, 153; units for,
59n10; vacuum, 179; various forms
of, 59; vibration, 95
entropy, 118, 181–82
enzyme, 183
equilibrium, 11, 12n15, *12*; static, 115

Flutie, Doug, 31–32, 37; "Hail Mary"
pass, 31–32, 31n2, 32n4, 41–44;
Heisman trophy, 32n4; trajectory of
"Hail Mary" pass, *42*, *43*
football: angular momentum and, 90;
drag coefficient of, 43n21; drag from
air, 43; ellipsoid, 43; gravitational
force on, 38–39; kicking, *23*; kickoff,
17n3; pressure in, 22, 22n11; stitches
on, *125*; structure of, 22; weight of,
38–39. *See also* soccer ball
force: Newton's first and second law
and, 38; Newton's third law and, 20,
21; mass and, 39n14; stress, 11; units
for, 20n10; vector nature of, 20. *See
also* friction; gravity, normal force;
spring
40-yard dash, 33, 33n5
Fosbury, Dick, 73
Fosbury Flop: 73; athlete executing *73*
free kick, 123, *124*; Beckham and, 121,
134–37; Carlos and, 144; parameter
space for successful, *138*; schematic
overhead view of, *135*; target for, *136*;
trajectory of, *139*
friction: air, 126, 134, 155; heating ice
and, 95–96; kinetic, 21–22, 95;
microscopic description of, 21–22;
rolling, 57; static, 21–22, 111, 186,
194; tires and, 159; torque and, 131
fulcrum, sketch of a, 104–5, *105*

Galilei, Galileo, 5–6, 5n9
geometry, use in terrain modeling, 51–52

glucose, 183–86, 190; mass of a mole of, 184n13
golf ball, 126
gravity: acceleration due to, 38–40; center of, 71–72

"Hail Mary" pass, 17, 31–32, 31n2, 32n4
helicopter, lift and, 166
honeybee, wing flap rate of, 4
hydrogen bond, 93–94
hydrolysis, 184, *189*

impulse, 192–93; baseball glove and, 193
inclined plane, for constructing terrain model, 52–55, 64n15
isotope, 93n2

kinematics, 33, 44, 106
Kosar, Bernie, 31–32

laminar flow, *127*, 128. *See also* soccer ball
lateral, in football, 16, 27–29; vectors needed to study, *28*
lift coefficient, 132n10, 166
light, speed of, 2–3, 6, 9, 14, 179
lightning, calculating distance of, 10
linear momentum: conservation of, 75, 194–95; definition of, 191–92; Newton's third law and, 20n8
Louganis, Greg, 91, 102–3, 107–9, *110*, 111, *112*, 113; dive trajectory of, *108*

Magnus, Heinrich Gustav, 132
Magnus force, 132–34, 163–64; reverse, 133n12
mass: center of, 71–73; Earth's, 77; moment of inertia and, 76; Newton's second law and, 38, 61; scalar nature of, 39; weight and, 39, 39n14, 54
mitochondria, 183, 185
moment of inertia: Beamon and, 84; changing, 81; definition of, 76–77; Earth's, 77; Louganis and, 109; tensor nature of, 76n6, 100n5; Witt and, 98, 100–102

muscle: abdominal, 85; agonist, 190; antagonist, 190; cell, 186–87; contraction and, 190; energy storage and, 154–55; oxygen release and, 186; *rikishi* gut and, 195; skeletal, 190; toy model of, *187*, *189*; weak, 116
myofibril, 187; toy model of, *187*, *188*
myosin filament, 188; toy model of, *188*
myosin head, 188, *189*, 190–91
Myron, *Discobolus*, 145–47

neutron, 8–9, 76–77, 93, 93n2, 184n13
Newton, Isaac, 9, 9n12, 132, 132n11, 192
Newton's first law, 14n17, 38, 157, 157n8; conceptual question (marble in tube), 156–57, *156*
Newton's second law, 38, 61, 71, 114, 132–33, 135, 137, 158, 168, 191–92, 195; calculus and, 192n18; rotational form of, 76, 80n14
Newton's third law, 20, 20n8, 22, 54, 54n5 105, 126, 131, 166, 194
normal force, 55, 57, 194
Nostradamus, 145n1

Oerter, Al, 146–51, *148*; art and, 172–73, 173n15; discus trajectory of, *173*; strength of, 158; throwing motion of, 154–56
optical measuring device, 68–69, *69*
oxidation, 183, 190

photosynthesis, 179, 182–83, 190
Powell, Mike, 89, 89n17
power: biking, 58–60, 62–64, 66; definition of, 58; stroke (muscle), *189*; units of, 60n11
Prandtl, Ludwig, 126, 127n5
pressure, 22n11, 115; fluid, 115; football and, 22; gauge, 22n11; ice and, 95–96
projectile motion, 37, 73; equations for, 40; trajectory in vacuum for, 42
proton, 8–9, 72, 76–77, 93, 184n13; sun, 179, 179n11

quantum mechanics, 9; tunneling, 179, 179n11